AutoUni – Schriftenreihe

Band 105

Reihe herausgegeben von/Edited by
Volkswagen Aktiengesellschaft
AutoUni

Die Volkswagen AutoUni bietet Wissenschaftlern und Promovierenden des Volkswagen Konzerns die Möglichkeit, ihre Forschungsergebnisse in Form von Monographien und Dissertationen im Rahmen der "AutoUni Schriftenreihe" kostenfrei zu veröffentlichen. Die AutoUni ist eine international tätige wissenschaftliche Einrichtung des Konzerns, die durch Forschung und Lehre aktuelles mobilitätsbezogenes Wissen auf Hochschulniveau erzeugt und vermittelt.

Die neun Institute der AutoUni decken das Fachwissen der unterschiedlichen Geschäftsbereiche ab, welches für den Erfolg des Volkswagen Konzerns unabdingbar ist. Im Fokus steht dabei die Schaffung und Verankerung von neuem Wissen und die Förderung des Wissensaustausches. Zusätzlich zu der fachlichen Weiterbildung und Vertiefung von Kompetenzen der Konzernangehörigen, fördert und unterstützt die AutoUni als Partner die Doktorandinnen und Doktoranden von Volkswagen auf ihrem Weg zu einer erfolgreichen Promotion durch vielfältige Angebote – die Veröffentlichung der Dissertationen ist eines davon. Über die Veröffentlichung in der AutoUni Schriftenreihe werden die Resultate nicht nur für alle Konzernangehörigen, sondern auch für die Öffentlichkeit zugänglich.

The Volkswagen AutoUni offers scientists and PhD students of the Volkswagen Group the opportunity to publish their scientific results as monographs or doctor's theses within the "AutoUni Schriftenreihe" free of cost. The AutoUni is an international scientific educational institution of the Volkswagen Group Academy, which produces and disseminates current mobility-related knowledge through its research and tailor-made further education courses. The AutoUni's nine institutes cover the expertise of the different business units, which is indispensable for the success of the Volkswagen Group. The focus lies on the creation, anchorage and transfer of knew knowledge.

In addition to the professional expert training and the development of specialized skills and knowledge of the Volkswagen Group members, the AutoUni supports and accompanies the PhD students on their way to successful graduation through a variety of offerings. The publication of the doctor's theses is one of such offers. The publication within the AutoUni Schriftenreihe makes the results accessible to all Volkswagen Group members as well as to the public.

Reihe herausgegeben von/Edited by
Volkswagen Aktiengesellschaft
AutoUni
Brieffach 1231
D-38436 Wolfsburg
http://www.autouni.de

More information about this series at http://www.springer.com/series/15136

Stephan Ahbe · Simon Weihofen
Steffen Wellge

The Ecological Scarcity Method for the European Union

A Volkswagen Research Initiative: Environmental Assessments

OPEN

Stephan Ahbe
Darmstadt, Germany

Steffen Wellge
Wolfsburg, Germany

Simon Weihofen
Essen, Germany

Any results, opinions and conclusions expressed in the AutoUni – Schriftenreihe are solely those of the author(s).

AutoUni – Schriftenreihe
ISBN 978-3-658-19505-2   ISBN 978-3-658-19506-9   (eBook)
https://doi.org/10.1007/978-3-658-19506-9

Library of Congress Control Number: 2017954959

This Springer imprint is published by Springer Nature
The registered company is Springer Fachmedien Wiesbaden GmbH
The registered company address is: Abraham-Lincoln-Str. 46, 65189 Wiesbaden, Germany

Acknowledgement

The authors and publishers would like to express their sincere thanks to the following members of staff of the **European Environment Agency** in Copenhagen for their committed, expert assistance in the compilation of the data regarding current conditions in the European Union and the corresponding future environmental targets: Almut Reichel, Katja Rosenbohm, Beate Werner, Martin Adams, Bo Jacobsen and Paul McAleavey.

Our thanks also go to Paolo Canfora and Harald Schönberger at the **Institute for Prospective Technology Studies** at the European Commission's Joint Research Center in Seville for the intensive discussions and the documents made available to us.

Dr. Stephan Ahbe
Dr. Simon Weihofen
Dr. Steffen Wellge

Preliminary Remarks

The present study "Ecological Scarcity Method for the countries of the European Union" is the successor to a similar investigation relating to the geographic region of Germany: "Ecological Scarcity Method for Germany" (Ahbe *et al.* 2014, Methode der ökologischen Knappheit für Deutschland). To facilitate comprehension, individual passages from the prior study which are of significance to both geographic regions or are otherwise of assistance in explaining the application of the method have been included in the present document. These passages are not formally cited except where the source is not obvious from the context.

Table of Contents

Table of Figures and Tables

1 Management Summary

The Ecological Scarcity Method (ESM) enables measurement and assessment of the environmental impacts caused, for example, by manufacturing sites. Developed in Switzerland in 1990, where it has since been in use, the method is constantly being developed and updated. It has already gained regulatory status in Switzerland, for example for the purpose of proving entitlement for tax exemptions, in particular for environmentally friendly production of bio-fuels[1]. A data set has been available for Germany since 2014. The method assesses all important environmental impacts on the air, on water, the consumption of energy, the generation of waste and the consumption of freshwater.

Assessments of this kind are necessary, in particular for manufacturing companies, for assessing the environmental impacts caused by economic activity. They may, for example, take the form of environmental impact assessments of manufacturing sites or even of individual production processes. Such assessments are also a tool for identifying which amount of capital expenditure in field will have the greatest effect on the environment. When it comes to answering these questions, the reliability and traceability of the assessment is of vital significance to the companies basing decisions on these results.

The name "Ecological Scarcity Method" is derived from the fact that the environment's capacity for pollutants is limited up to a critical state, i.e. is scarce.

In the ESM, the scarcity situation is defined by the current, existing environmental impact and the capacity, as defined by a country's highest environmental authorities, to withstand this impact as a target state. This ensures that the various users of the assessment method will make use of a common basis of assessment, so making the assessment user-neutral, objective and reproducible at any time.

The scarcity situation of the environment with regard to a pollutant thus depends on the difference between the current environmental impact, for instance in tonnes of pollutant per year, and the quantity or "critical environmental impact" which, on the basis of the environmental objectives, is still just about acceptable. Each pollutant discharge or also each consumption of resources takes place against the background of a corresponding scarcity situation. The consequent relative deterioration in the scarcity situation, a ratio, can be added up for all such impacts, giving rise to the total environmental impact, for example for a manufacturing site in a specific year.

Before the assessment method can be applied, it is vital for the most important environmental impacts for the country under consideration to have been investigated as completely as possible. Such is the case in the EU and also in many other countries.

---

1   cf.: Swiss Confederation "Fuel life cycle assessment ordinance" of 9.4.2009, paragraph 6.

One significant feature of the ESM is that it can be used to assess completely different environmental impacts and to compare them with one another. The feature common to all environmental impacts, which makes such comparability possible, is the relative deterioration in the scarcity situation brought about by each individual environmental impact.

Overview 1:   The present study researched eco-factor data for the following countries

EU 28 (regarded as one environmentally decision-making unit)

Austria
Belgium
Bulgaria
Croatia
Cyprus
Czech Republic
Denmark
Estonia
Finland
France
Germany (for the purpose of comparison)
Greece
Hungary
Ireland
Italy
Latvia
Lithuania
Luxembourg
Malta
Netherlands
Poland
Portugal
Romania
Slovakia
Slovenia
Spain
Sweden
United Kingdom

The data for each country is available in chapter 6.

The method allows a user directly to establish whether the overall environmental impact on a site has fallen or risen and which individual impact has had what influence even in the presence of countervailing trends.

Using the ESM, environmental impacts may be calculated and classified in the form of eco-points (EP) and also used in various management tools for setting objectives, in a similar way to that known from management cost accounting.

The present report from SYRCON Darmstadt describes the transfer of the method to European conditions with a survey of the corresponding current impacts and the target impacts drawn up and published by the competent environmental authorities, specifically the European Environment Agency.

2 Introduction

2.1 Method Description

The "Ecological Scarcity Method" (ESM) was developed to make the environmental impacts, which arise during the everyday commercial operation of manufacturing sites or plants, measurable, assessable and comparable. The method was developed by industry in Switzerland between 1987 and 1990 because no reliable assessment method was available but industry and commerce were nevertheless increasingly being expected to address environmental issues. The method has been used ever since. It has constantly been developed, and continously been further and kept completely up to date with regard to the basis for assessment. In the meantime, it has already gained regulatory status, for example being specified for the purpose of proving entitlement to tax exemptions for particularly environmentally friendly manufacturing, for instance in the production of biofuels. The method assesses all environmental impacts that are considered significant by the environmental authorities, including emissions to the air and surface water, consumption of energy, freshwater and waste generation.

Such an assessment is required, for example, in order to establish whether a manufacturing plant has reduced its overall environmental impact compared to the previous year. It can also be used to answer the question as to which investments in improvements to production facilities will achieve the greatest reduction in environmental impact or what measures can be implemented for a given capital expenditure to achieve the greatest environmental benefit. Being able to answer these questions reliably and traceably and to derive appropriate environmental targets from the answer is of vital significance to entrepreneurial decision making.

The name "Ecological Scarcity Method" was selected because the environment only has a limited capacity to take up pollutants before an intolerable state is reached. In other words, the environment's capacity for pollutants is "scarce" by analogy the same applies to the availability of resources.

In order to describe this scarcity situation as accurately and traceably as possible, the ESM uses the environmental targets set by a country's or geographic region's highest environmental authorities. This approach is intended to ensure that everyone applying the assessment method will use the same data basis, and thus, the same environmental goals for assessment, thereby ensuring that different assessors do not obtain different results for the same situation.

The scarcity situation of the environment with regard to a pollutant thus depends on the difference between the current environmental impact, for instance in tonnes of pollutant per year, and the quantity or "critical environmental impact" which, on the basis of the environmental objectives, is still just about acceptable. Each pollutant

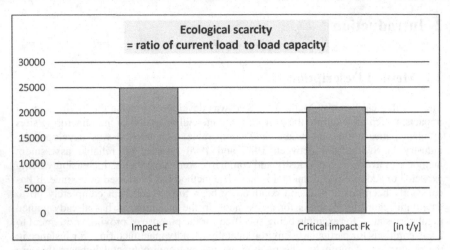

Figure 1: Principle of ecological scarcity

discharge and each consumption of resources takes place against the background of a corresponding scarcity situation. The consequent relative deterioration in the scarcity situation, a ratio, can be added up for all such impacts, giving rise to the total environmental impact, e.g. for a manufacturing site over a specific period.

Before the assessment method can be applied, it is vital for the most important environmental impacts for the country in question, to have been investigated and quantified as completely as possible. Such is the case in many countries and also in the EU as a community of nations. One significant feature of the ESM is that it can be used to assess completely different environmental impacts and to compare them with one another. In a similar way to how, say, apples and pears can be compared with one another, for instance by a freight carrier, because they share a common feature. That is for example their weight in kilograms, which is of importance to the carrier. The feature common to all environmental impacts which enables this comparability is the relative deterioration in the scarcity situation brought about by the particular individual environmental impact under consideration.

A user of the method can then tell directly from the result whether, for example, a site has become "more environmentally friendly" over the course of a year, despite energy consumption having perhaps increased, waste volumes having fallen, greenhouse gas volumes having risen and emissions to water having declined, etc.

Using the ESM, it is possible to environmental impacts, prioritize a budget and define environmental objectives. In brief, the structures for corporate use are largely comparable with those of management cost accounting (Please note that the calculation process used for the ESM is described in greater detail at the beginning of the data section).

2.2 Initial Situation and Aim of the Initiative

The aim of this initiative is to create the basic conditions under which European companies will be able to draw up meaningful, reliable environmental impact assessments. When taking environmentally significant corporate decisions, it is essential for the assessment results to be traceable by third parties. The results must be free of subjective considerations, i.e. they cannot be dependent on the assessor. This is apparent when it is borne in mind that considerable financial resources are often invested in environmental matters; consequently a traceable and reliable basis for decision making must be created. Similarly, environmental management systems, such as EMAS or ISO 14001, require decision making tools for the assessments which permit quantification and comparability.

As with the transfer of the ESM to Germany, Volkswagen AG has again taken initiative with the aim of creating the necessary conditions for a reliable and comparable basis for environmental impact asssements across Europe. The ESM has already been available as a basis for assessment of Swiss conditions for a long time. Since 2014 it is also available for German conditions too, the emphasis here has been placed on surveying, processing and coordinating the targets of European environmental policy. In autumn 2010, SYRCON held an industry seminar on this issue in Darmstadt. Interested companies were welcome to obtain information about the state of environmental impact assessment methods in general and the ESM in particular. The Swiss Federal Office for the Environment also reported on positive practical experience with the ESM in Switzerland on the basis of specific case studies. Technische Universität Darmstadt presented a comparative investigation of environmental impact assessment systems which demonstrated that the ESM was fundamentally suitable for corporate use in Europe, also in the context of environmental management systems. A joint concluding statement from the seminar participants emphasised that, in the light of the clear need for a method of this kind, it would be logical and desirable to transpose the ESM initially to German and then to European conditions as soon as possible.

In December 2014, as part of an initiative by Volkswagen AG, the "Ecological Scarcity Method" was transposed to German conditions as an environmental impact assessment method and German Eco Factors were published (Ahbe *et al.* 2014). By publishing the data, Volkswagen AG made the German Eco Factors available for public use and also for other companies wishing to engage in active environmental management at their sites. In most cases, these business activities are not limited to Germany but extend across Europe or worldwide. This predicates the determination of Eco Factors for the member states of the European Union, as described here. Determining Eco Factors for the member states of the European Union and further selected countries worldwide also forms part of Volkswagen's initiative.

2.3 Objectives for European Data

With regard to environmental policy, the European Union (EU) should be considered
to be a geographic unit in which targets for the state of the environment and measures
for achieving these objectives are defined and implemented. The Lisbon "Treaty on
the functioning of the European Union" of December 2007, specifically articles 191–
193, sets out the fundamental principles of common environmental policy.

It should be emphasised that the goals of European environmental policy were drawn
up independently of any possible use in this or another environmental impact assessment method and that they can equally be applied for other purposes and methods.

Thus, the basic conditions for determining Eco Factors for the European Union as a
geographic region with its own environmental policy and own objectives are in place.
It should be noted, however, that differences in terms of objectives most certainly exist
between EU environmental policy and environmental policies at the level of the
individual member states. It may be that national government targets are stricter or
laxer than those at EU level.

What may at first sight appear to be an inconsistency is entirely understandable on
closer examination: the assessment method is guided by the applicable geographic
affiliation and the resultant environmental policy objectives that apply. Competition
between targets at a national government level and at an EU level certainly occurs and
results in different basic conditions for calculating the Eco Factors, depending on the
geographic and political context.

When transferring the Ecological Scarcity Method to EU member states, depending on
location, differing normalisation flows, current flows and critical flows must be
expected in comparison with the previous data sets for Switzerland and Germany.

Critical flows are synonymous with environmental policy objectives. The different
political circumstances in Switzerland, Germany and the EU are an important factor.
In line with instructions, the environmental policy objectives of the EU, where these
exist, are used as the benchmark for assessment.

In view of its specific remit, the European Environment Agency (EEA) has also made
a substantial contribution to determining Eco Factors for EU member states in the
present paper. Thanks to wide-ranging documentation and ongoing data exchange with
individual EU countries, robust data sets have been obtained for use in the Ecological
Scarcity Method which takes proper account of the necessary neutrality and independence of any particular industry.

The present paper largely refers to EEA with its environmental information and
stipulations. Where gaps in data sets were identified or no geographically appropriate
targets have been set, an attempt is made to find a solution by suitable approximation
methods. Such cases are explicitly stated. Crucially, all interpolations or extrapolations
and other determinations take their lead as closely as possible from the stated political

will of the environmental authorities and are free from any subjective influences by those carrying out the study.

The companies that base their environmental impact assessments on the data from this study have an associated expectation that the relevant official bodies will, to the best of their ability, maintain, update, revise and extend the available data set. It would be desirable for greater application of the present data also to lead to an increased readiness by all concerned players to update the data periodically and adapt it to the most recent circumstances.

2.4 Project Implementation

The "Ecological Scarcity Method for the European Union" project is an initiative of Volkswagen AG, which commissioned the engineering consultancy SYRCON Dr. Ahbe, Dr. Popp & Partner in Darmstadt to carry out the project. The project ran from July 2013 until March 2015.

3 Methodological Bases

3.1 Ecological Scarcity Method

A detailed description of the method with explanations regarding the calculations can be found in (Frischknecht *et al.* 2013) and (Ahbe *et al.* 1990, 2014) as well as in chapter 7.

3.1.1   How does the ESM fit to the Phases of the ISO Standard?

According to ISO standards 14040:2006 and 14044:2006, an environmental impact assessment is divided into four phases:

1) Goal and scope definition
2) Life cycle inventory analysis
3) Impact assessment
4) Interpretation

Using these ISO phases as a basis, the ESM, in accordance with its objectives, involves a method for impact assessment and interpretation once a life cycle inventory analysis has been correctly drawn up. The latter, together with a correct scope definition, have already been described in detail in the relevant literature (cf. Frischknecht *et al.*, 2013, ISO 2006).

3.1.2   What are the Elements of the ESM?

The Ecological Scarcity Method essentially consists of three elements:

1) the assessment and aggregation algorithm as calculation specification (always remains the same),
2) the data set for the target country (in this case the EU), consisting of the selection of environmental impacts to be taken into consideration with the associated current impact values and the quantitative target environmental impacts which are described by the environmental authorities and should, if possible, be outperformed,
3) the life cycle inventories, for instance of the company's site or processes under investigation, which are to be assessed.

3.2 Basic Principle

3.2.1   How can the ESM be applied?

The ESM can be applied anywhere in which different environmental impacts need to be meaningfully assessed and compared with one another. One possible focus in the corporate field is the assessment of manufacturing sites or of comparable process steps which need to have their environmental impact reduced. Another widespread application is the analysis of different possible manufacturing scenarios with corresponding derivation of achievable environmental targets.

The internal cohesion between these forms of assessment is that, for instance, a production site may be viewed as an aggregate of the processes that take place on site and the assessment of these processes is conducted within one plant (see Figure 2). Further applications can be found in Frischknecht *et al*. 2013.

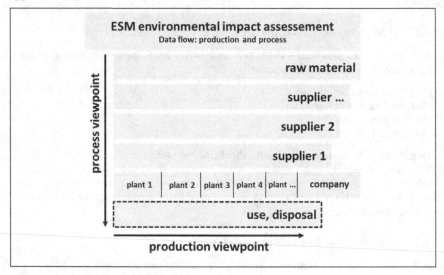

Figure 2:   Data flow in ESM environmental impact assessment

3.2.2   *Which Requirement apply to the ESM and its underlying Data?*

In environmental impact assessments, particularly if carried out for the purpose of corporate presentation, care must be taken to avoid creating any impression of self-interested bias in the results. Years of prior industrial experience with handling environmental impact assessments led, from the initial design stage for the ESM back in 1990, to specification of the following requirements that are indispensable for achieving the desired transparency and that also apply to the treatment of data in the present paper:

Completeness

The assessment must include all substantial environmental impacts of the process or site under consideration.

Public goals

The assessment must be made on the basis of the published environmental goals of the relevant national authority and, for reasons of neutrality and traceability, these must be used as benchmarks for the assessment.

Independence of the author of the assessment

The results of the environmental impact assessment must be independent of the author of the assessment (as in the drawing up of business accounts).

Unambiguous statement

When making comparisons, the statements made in the assessment must be unambiguous and be usable and reproducible in business practice.

Systematic aggregation

Aggregation of the assessment result to form an overall statement must be systematic and must not be left to the discretion of a (subjective) user.

3.2.3 Coordination with Environmental Authorities

One essential principle of the ESM is to coordinate data sets with the relevant highest environmental authorities with responsibility for a particular geographic or policy area. This ensures that both the actual state of the environment and the desirable nominal state concur with any existing determinations made by the regulatory authorities, thereby avoiding any bias in terms of viewpoint or assessment by the author of the environmental impact assessments. Another significant reason for this is that these assessment data should be identical for all parties drawing up environmental impact assessments. Having an identical and uniform basis for assessment is an absolutely necessary prerequisite for communicating the results to third parties.

3.2.4 Requirements for European Eco Factors

The Eco Factors for each environmental impact are calculated in accordance with the method's calculation specification essentially from the two loads or consumption variables that substantially determine the particular environmental situation:

■ current annual flow (or consumption), synonymous with current annual load,
■ critical annual flow (or consumption), synonymous with a target annual load, exceeding which can no longer be considered acceptable.

Current flow can be determined from previous surveys and statistics compiled to describe the actual state of the environment with regard to the particular impacts under consideration in the country in question.

Critical flow, on the other hand, can only be determined once a prior environmental policy decision has been made defining the 'just about acceptable' state of the environment. By definition, this is a per impact environmental objective which is articulated at European Union level by the relevant highest authorities and implemented in the respective countries in the context of plans of action. The administrative body with competence for setting environmental targets is the European Commission with the relevant Directorates-General as the executive of the EU, together with their relevant institutes and subordinate authorities, the most important of which are listed below:

European Environment Agency in Copenhagen

In line with its remit, the European Environment Agency (EEA) in Copenhagen, Denmark, plays a prominent role in the present case. The purpose of the EEA is, in its own words (EEA 2009):

> "The task of our agency, which has some 130 staff and an annual budget of €40 million, is to contribute to shaping European and national policy by providing independent information and assessments on environmental issues. Our work focuses on the following areas:
>
> – state of the environment;
> – current developments, including impact of economic and social factors;
> – policy strategies and their effectiveness;
> – possible future trends and problems."

The European Environment Agency is thus a key interface between the environmental authorities of each EU member state, acting as a coordinating body for the individual countries' environment status reports and, in this role, also supplying neutral environmental information to interested parties. The EEA supplied the basic data for the present study.

Institute for Prospective Technological Studies (IPTS) in Sevilla

IPTS in Sevilla, Spain, is part of the Joint Research Center of the European Commission. It concentrates on "best-practise" questions in the field of technology as well as on possible political measures for a target-oriented controlling of corresponding technological developments. With members of IPTS conversations were hold and informations were exchanged in this study.

Institute for Environment and Sustainability (IES) in Ispra

IES in Ispra, Italy, is likewise part of the European Commission Joint Research Centre. Its main mission is to research methods for recording environmental impacts and their consequences. Its activities therefore focus on the possibilities for using environment-related data and assessment methods. IES is mentioned for the sake of completeness, but had no direct influence on the content and course of the present study, since the assessment method itself was not at issue but instead the transposition of the method to other geographic regions.

3.3 Method

3.3.1   Requirements for European Eco Factors

The ESM enables comparing different environmental impacts with one another and to convert them into a single aggregated value. In simple terms, the aggregation mechanism states the "degree of undesirability" of an environmental impact. This is characterised by the ratio of the current state of the environment to the desired mitigation target, i.e. the scarcity situation. A level of undesirability exists for each and every environmental impact in a given region or country, making this a criterion common to all impacts that is sufficiently meaningful for the purpose of aggregation.

3.3.2   What further Options are there for Applying the ESM?

Because the individual environmental impacts are directly comparable, the corresponding eco-points (EP) can serve as a unit of measurement for environmental impacts and be used in various forms for the findings of the assessment (see data section for more details about the calculation formula). Accordingly, such direct comparisons may be used, for example:

1) to rank different investment options,
2) to draw up environmental budgets, for instance per site, per sector or department, or
3) to determine environmental impact, for instance per tonne of manufactured product, per unit or the like, and
4) to define measures to maximise environmental mitigation within a specified period or financial budget and
5) for further *ad hoc* purposes.
6) Financial management accounting can provide a point of reference for the basic structure of possible assessment approaches and the use of eco-points.

3.3.3   How can Traceability be communicated?

The traceability of environmental impact assessments is of vital significance especially when communicating with third parties such as clients, competitors, industry associations, authorities, auditors etc.. Were any doubts arise in this connection, the assessment result would be considerably devalued and its suitability may be as a basis for decision making called into question. To ensure a high level of reliability, the assessment result must be free of subjective considerations on the part of the assessor, since other assessors would otherwise come to different assessment results. It would be counterproductive for business decision making if environmental decisions could be undermined in this respect. The logical comparison is with drawing up a corporate balance sheet: the balance sheet total must be totally independent of the person drawing it up, if banks, creditors and investors are to be able to rely on a disinterested, neutral picture of a company's current financial circumstances.

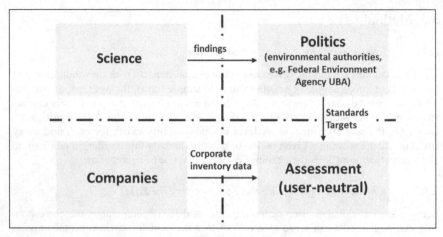

Figure 3: "Separation of powers" in the ESM, taking Germany as an example

Figure 3 shows how bias and arbitrariness are largely eliminated from the assessment result by applying a "separation of powers" in the ESM.

The essential factor is that there is a substantive and personal separation not only between obtaining the scientific findings on the one hand and the setting of standards and targets by the environmental authorities on the other, but also from the company which, while indeed generating and collecting environmentally relevant life cycle inventory data arising from its business activities, has no influence that could affect the assessment on the first two processes and so cannot influence the result by selecting the assessment parameters. Third parties would doubtless suspect bias if the company drawing up the assessment also wished to define the problematic environmental impacts to be assessed and quite possibly the associated environmental targets as well.

3.3.4   What Transparency Rules need to be observed?

One significant characteristic of the ESM is that the assessment result does not depend on the particular assessor, providing that some important rules are observed:

1) Role of the authorities: the data sets used for the environmental impact assessments have a decisive influence on the assessment results. It is therefore crucial for reliable communication both within and outside a company that the basis for assessment is free of any possible self-interest on the part of the company. This is ensured by using the targets officially set by the highest environmental authorities for determining ecological scarcity values. This avoids any impairment of the credibility of the assessment result due to suspected bias on the part of the assessor.

2) Statement of sources: when environmental data are used, it must always be evident which data have been used and where they come from. The data set used (for EU,

Germany, Switzerland etc.) must always be clearly stated in the assessment. The data sets for different geographic regions are not interchangeable, since both the respective national loads and the corresponding environmental targets differ from one country to the next. Within an assessment, the data set used must be clearly stated and always used in any direct comparison.

3) Date of publication of the data set: this must likewise be stated because the update status within a geographic region must be known. The reason for this is obvious: both the scarcity situation of the individual environmental impacts and the extent of environmental impacts under consideration can change over time. For Switzerland, for instance, the fourth updated data set is already available (from 1990, 1998, 2006 and 2013). Another rule is always to use the most recent and thus most up to date data set.

4) Description of the data set: this should be as unambiguous as possible, for instance „ESM-D-2014" or "ESM-CH-2013" etc.

5) Comparability: if comparative assessments are to be performed, care must be taken to ensure that the investigated variants are actually comparable. Accordingly, one production process can only be compared with another if it leads to the same output. Or a manufacturing site can be compared over two periods (e.g. financial years) in order to determine whether the total environmental impact has perhaps risen or fallen over time. Similar criteria for comparability are also found in management cost accounting.

6) Data accuracy: one inherent aspect of environmental impact assessments is that it is rarely possible to obtain completely accurate data sets and so it is often necessary to work with averaged, interpolated or extrapolated values. The relative assessment error can frequently also be reduced by working with lower accuracy data if the alternative is to omit the corresponding data completely because of concerns about inadequate accuracy. This should always be declared, especially in contact with third parties.

7) Data updating: the data sets used for assessment should be periodically updated. A period of 5 to 7 years has proved to be appropriate, after which the underlying data are checked or adapted to the new circumstances. Obviously, new findings may have been made in the meantime concerning the extent of the environmental impacts under consideration and the respective current and critical impacts. Updating at shorter intervals makes less sense because it results in numerous versions of data sets which differ only slightly from one another and are consequently unsuitable for comparison.

3.3.5   Basis for Assessment

The basis for assessment which underlies the method is of particular significance. In the Ecological Scarcity Method, this basis is provided by the environmental policy defined by the authorities for desirable targets by level and by the time horizon until these targets are achieved. The environmental policies of the individual countries are essentially in competition with those at EU-28 level, since, from a purely geographic standpoint, both are valid for any location within the European Union.

EU-28 with environmental policy management

European Union environmental policy has become considerably more influential and significant in recent years. This is because the scope and diversity of EU programmes have substantially increased in content terms over recent years. In policy terms, there is a discernible trend for individual countries to hand over more and more responsibilities to the EU. As a consequence, a wide-ranging environmental programme backed by policy targets is now in place at EU-28 level. The European Commission acts autonomously here, making its own determinations regarding the future state of the European environment. The statistical surveys required to describe the current environmental situation are likewise carried out on an extensive scale. Thus the environmental policy of the European Commission can be used as a neutral standard of evaluation specifically for the communication with third parties like authorities, customers, competitors etc.

Individual country data as an interim solution

Within Europe, eco-factor data sets for the ESM assessment method are currently available for Switzerland and Germany. In Switzerland, this data set has been maintained for many years and is constantly being extended. In Germany, Eco Factors were first determined in 2014 and coordinated with German environmental policy targets with the collaboration of the Federal Environment Agency (UBA). The requirements of the ESM are thus met in both cases.

The Eco Factors which were derived by calculation from EU data and used in the present paper for the **27 individual EU countries** other than Germany meet these requirements only in part. One essential fact to be borne in mind is that no coordination with the highest environmental authorities of the individual countries has yet taken place and the countries have thus not formally identified with the content of the underlying targets. However, in line with the strict requirements of the Ecological Scarcity Method, this is an essential prerequisite for the unreserved acceptance of the assessment results by third parties, some of whom have very different basic attitudes towards environmental issues.

For the Eco Factors for the individual countries to be obtained deductively as described here from existing EU data in order to save time and effort can be considered a first major step forwards. For reasons of acceptance, it would be important for coordination with each country's specific environmental policy to be achieved over time. This way, the aim of providing a neutral basis for assessment which is accepted by all parties and takes account of each country's environmental targets will remain in focus.

3.3.6   Rules for Assessment

A number of general rules can be defined for environmental impact assessments:

Assess as directly as possible

In other words, the more specific, regional – i.e. in this case the national governments – Eco Factors should normally be used if they have been properly determined in accordance with the above methodology and have been coordinated with the aims of the respective country's environmental authority.

Taking a wider view

The following circumstances may necessitate assessment on the larger scale of EU-28 environmental policy as an exception to the above rule:

- if different sites or processes in a number of European countries are to be compared or aggregated, as may for example be the case for company-wide investigations across various manufacturing sites, or
- if no or too few Eco Factors are available for an individual country, so that the only Eco Factors available in sufficient quality and quantity are those from the next higher geographic level.

In both cases, it makes sense to carry out an European assessment using EU-28 Eco Factors covering the whole of Europe. It is therefore absolutely essential for the documentation to state precisely which data set has been used for the assessment, since the result is, as described above, numerically dependent on this (cf. Ahbe *et al.* 2014).

3.3.7   What must be borne in Mind when drawing up Assessment?

Drawing up traceable, reliable environmental impact assessments which are suitable as the basis for decision making for major investments or for communication with third parties depends on a number of factors:

1) a carefully drawn up environmental inventory: this includes carrying out appropriate substantive analyses of the processes while taking proper account of issues such as handling of co-products, credit for recycled fractions, use of manufacturing waste, usage data, application of allocation rules and many others.
2) proper declaration of the data sets and sources used and assumptions made etc., thus ensuring transparency of the relevant basis for calculation for the assessment.
3) the name of the author of the assessment and, if applicable, the software and databases used must be stated to ensure traceability.

3.4 Methods

3.4.1   The ESM: Midpoint or Endpoint Method?

The scientific literature contains numerous systems, sets of rules and standards which are intended to be suitable for providing an overview of the various assessment methods and for assisting in informed decision-making. It makes sense in this connection to give proper consideration to all the elements of the assessment system so that traceable conclusions can be drawn.

The literature conventionally divides assessment methods into "midpoint" and "endpoint" methods. The former are those which, during assessment, focus on various impact categories such as ozone depletion, acidification, greenhouse effect etc. (midpoint indicators) while the latter are those guided by harm categories such as "human health", "biodiversity" etc. (endpoint indicators). The two types of method generally assume different horizons with regard to the range of the assessment, since endpoint methods include a further aggregation step and so are more likely to achieve the actual aim of assessing actual environmental harm.

Closer examination of the true situation reveals that this division and the associated interpretation lose their significance in the ESM. Since the official environmental targets used are unambiguously directed towards environmental harm and the deliberate avoidance of such harm, assessment is provided right up to an "endpoint", although the literature usually categorises this method in the midpoint category. In this case the answer to the above question depends on the nature and manner of determination of the official targets defined and not on the assessment method itself. In this respect, a clear distinction must be drawn between the assessment method and the data sets, which are independent of the method, stating the environmental targets (cf. Figure 4, see next page).

3.4.2   Does the ESM comply with ISO 14040:2006 and 14044:2006?

DIN EN ISO 14040:2006 describes the basic procedure and broad framework within which an environmental impact assessment (also known as a "life cycle assessment") is drawn up and provides information about the four phases of a life cycle assessment (goal definition, life cycle inventory, impact assessment, interpretation), reporting and critical review.

A life cycle assessment should in principle preferably be drawn up on the basis of scientific findings. Alternatively, use may be made of both further scientific approaches and international agreements. If none of these approaches is appropriate, decisions may be made on the basis of values which must then be described in detail (cf. Frischknecht et al. 2013).

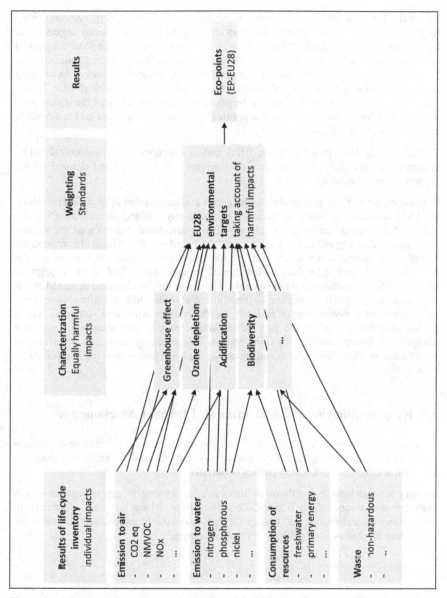

Figure 4: Alignment of environmental targets with harmful impacts (as described by the Swiss Federal Office for the Environment, Berne)

Scientific findings are only capable of revealing environmentally relevant inter-relationships. They do not in themselves set any targets. If environmental targets are to be defined, the state deemed "critical" with the associated annual load (or annual consumption) must be defined in the context of a consensus-building process. This process, where societally supported and guided along official pathways, becomes a societal or political declaration of intent, which then defines the critical state. In this way, a country's environmental policy target setting process includes the opinions of numerous stakeholders and is also supported by experts who then add a scientific dimension to the discussions.

As has already been mentioned, the ESM makes reference to environmental policy targets which are derived from scientific findings (cf. figure 3) and defined by the highest environmental authorities.

Because of the method used to obtain the targets and the applied aggregation principle, the ESM conceptually and systematically excludes any "subjective weighting" by the author of the assessment which might result in user-related distortion of the results. The aggregation algorithm used here is solely based on the official determinations regarding the actual and nominal state of the environment. In every assessment, the contribution of each individual environmental impact to the final result is quantitatively visible, countable and thus traceable. Thanks to this mechanism, the ESM meets two requirements: firstly, complete aggregation of all individual statements as required by industry as a system-supported basis for decision making and secondly, fundamental absence of bias on the part of the author of the assessment and thus utter transparency and traceability for third parties. It is precisely these two requirements from business which led to the development of the ESM assessment method (cf. Ahbe *et al.* 1990).

3.5 Responsible Use of Environmental Impact Assessments

If the above rules are consistently observed, the ESM can be used to draw up assessments for both in-house (internal) corporate requirements and for marketing purposes and comparative studies for third parties (external).

Industry associations and authorities also have an interest in assessments having a basis for assessment which is traceable at any time. Many years of experience in Switzerland have demonstrated that credibility is distinctly increased if the basis for assessment originates from publicly controlled sources.

3.6 Use of Data

3.6.1   Types of Impact under Consideration

From a corporate standpoint, the environmental impacts to be recorded with the ESM must be

■ known,
■ researched,
■ permitted and
■ planned.

Environmental impacts which are as yet unknown or those which have not yet been sufficiently well researched for it to be possible to derive corresponding environmental targets are accordingly excluded from assessment. Similarly, only permitted impacts are included since prohibitions, such as for example emission prohibitions, cannot be effectively implemented by means of such an assessment method. The method furthermore focusses on planned environmental impacts, i.e. those which are part of the process under consideration, and not for instance merely on existing risks for spills or other losses of process control.

Numerous anthropogenic impacts which influence the state of the environment are currently known. In the present project, Eco Factors were determined for the individual countries of the European Union on the basis of the same environmental impacts as for the preceding determination of Eco Factors for Germany. Selection criteria were:

■ the impacts are typical and significant in connection with business processes.
■ consultation with the German Federal Environment Agency regarding the impacts officially considered significant in this connection.

The types of impact under consideration are therefore also in the present case (cf. Ahbe *et al.* 2014) as follows:

■ air pollution by:
 – greenhouse gases as CO_2eq
 – NMVOC
 – NO_x as NO_2
 – SO_2
 – fine particulate matter (PM2.5)
 – NH_3

■ surface water pollution by:
 – nitrogen
 – phosphorus
 – nickel
 – zinc

- COD
- lead
- cadmium
- copper
- EPA-PAH 16

■ consumption of resources by:
- freshwater consumption
- renewable energy consumption
- non-renewable primary energy consumption

■ waste:
- waste generation, non-hazardous
- waste generation, hazardous

The decision initially to use the same types of impacts at an European level as were used for determining the German Eco Factors was made on the basis of the largely identical business processes and very similar weighting by the relevant authorities, some of which are also involved in joint reduction programmes.

From the standpoint of a multinational industrial concern, it is easier to implement environmental protection measures if a uniform scope of impacts is taken into consideration. This continues to apply until more recent findings preclude such a definition.

3.6.2 Comparability of Eco Factors and Eco Points

It follows directly from the rules for determining an individual country's Eco Factors that country A's aggregated assessment results cannot be compared directly and in absolute terms with country B's. This follows directly from the fact that the two countries are generally of different sizes and, not least for this reason, usually have different current annual flows and different quantitative environmental objectives, i.e. critical flows, which define the magnitude of the Eco Factors by calculation.

It is therefore absolutely necessary for the documentation to state precisely which data set has been used for a completed assessment, since the assessment result is, as described above, numerically dependent on it (cf. Ahbe *et al.* 2014).

The question frequently arises as to how an individual eco-factor, i.e. a scarcity state, per type of impact can adequately represent a large geographic region in which there may quite obviously be very different local scarcities. In other words, a production site may be located on a relatively unpolluted river as the receiving stream or alternatively on a river which is already highly polluted. Similarly, a production site emitting large volumes of air pollutants may be located in a relatively unpolluted, rural region or alternatively in a highly industrialized conurbation which is already highly polluted. Similar considerations apply to other types of impact.

For the purposes of scarcity definition, the ESM by definition focuses on the average conditions of a country or region. The emission and consumption data are here based on pre-existing impact patterns specific to this country which are determined by conurbations, industrial regions, surface water structure etc. and have given rise to the definition of the corresponding official environmental targets[1]. Against this background, if official targets exist which, for the purposes of refinement for instance, make reference to individual sub-regions of a country, these more detailed targets may also be used with the ESM.

A differentiation must be made here with site specific environmental impact assessments (EIA), which are drawn up directly for a specific location and are intended to assess the direct interaction between the emitters in question and their immediate surroundings[2]. If this is the intention, the relevant regulations for investigating and drawing up this site-specific EIA[3] must be observed.

---

1 Conditions at the point of discharge itself are often not critical, but instead the conditions arising from the total loads at a considerable distance from the point of discharge, for example in the North Sea, Baltic Sea or North Atlantic catchment basins.
2 This differentiation of EIAs also applies, due to their local focus, to other assessment methods.
3 For example the German Law on Environmental Impact Assessments (UVPG).

4 Data Collection

4.1 Methodology

According to (Frischknecht, Steiner, & Jungbluth, 2009), the Ecological Scarcity Method may be described as follows:

The Ecological Scarcity Method weights environmental impacts, i.e. pollutant emissions and resource extractions, with what are known as "Eco Factors". The Eco-Factor is derived from environmental legislation or corresponding policy targets. In its basic form, it follows the procedure of DIN EN ISO 14040. The eco-factor is defined as follows for each environmental impact:

Formula 1: Calculation of an eco-factor

$$eco - factor = \underbrace{K}_{\substack{\text{Characterization} \\ \text{(optional)}}} \cdot \underbrace{\frac{1 \cdot UBP}{F_n}}_{\text{Normalisation}} \cdot \underbrace{\left(\frac{F}{F_k}\right)^2}_{\text{Weighting}} \cdot \underbrace{c}_{\text{Constant}}$$

where:
- K = characterization factor of a pollutant or resource;
- F_n = normalisation flow: current annual flow, relative to the respective country;
- F = current flow: current annual flow, relative to the reference area;
- F_k = critical flow: critical annual flow, relative to the reference area;
- c = constant (10^{12}/y): ensures that numerical values that are easy to represent are obtained;
- EP = eco-point: unit for the assessed environmental impact.

Characterization factors are determined for pollutants and resources which can be associated with a specific environmental impact (for example the greenhouse effect). This involves relating the effect of a specific pollutant (for example the greenhouse effect of methane) to the effect of a reference substance (in this example carbon dioxide).

The purpose of **normalisation** is to adapt the scarcity situation (weighting) to current emissions/resource extractions in a region. Normalisation adapts (normalises) the evaluation to national conditions. Normalisation is therefore performed on the basis of the entire annual pollutant emissions/resource extractions of the country in question.

The final **weighting** of pollutants/resources or of characterised environmental impacts is performed on the basis of their "ratio to the environmental objective" or "**ecological scarcity**". This is achieved by relating the entire present flows of an environmental impact (current flows) to the maximum flows of the same environmental impact which are considered admissible for the purposes of environmental policy targets (critical flows). Depending on the nature of the environmental target or environmental legislation, this is carried out either on the basis of individual substances (environmental impacts) or (characterised) environmental impacts.

Scarcity (weighting) is a dimensionless quantity solely determined by the ratio of current to critical flow. The absolute magnitude of the flows has no influence on the weighting.

Factor c is identical for all Eco Factors and serves to simplify presentation, permitting a more manageable order of magnitude and including the time dimension which remains from the quantity units and to which it relates on a magnitude basis.

Being the result of characterization, normalisation and weighting, the **Eco Factors** represent the political and legislative assessment of the ecological significance of the pollutants. The unit for the eco-factor is "eco-points (EP) per unit quantity", for example "EP/g SO_2". Multiplying the eco-factor by the quantity of emission or consumption provides the eco-points per impact which are aggregated, i.e. summed, across all impacts.

Formula 2: Calculation of eco-points

$$EP = eco - factor * quantity$$

Some Eco Factors may be derived in a number of different ways. For the Ecological Scarcity Method, the principle is that in each case the highest of the resultant Eco Factors is used. Weighting is thus performed on the basis of the dominant assessable environmental impacts.

4.2 Principles for Deriving Eco Factors

The following principles for deriving Eco Factors are taken from the Swiss Federal Office for the Environment publication on the ESM (Frischknecht, Steiner, & Jungbluth, 2009).

4.3 Use for Characterization Factors

The fundamental condition for using characterization factors is that the characterization should be in line with the legislative intent.

In addition:

a) the characterization factors used should be scientifically recognised;
b) characterization factors may be derived from policy objectives.

4.3.1   Determination of Normalisation

The current flows used for weighting are generally identical to the flows to be used for normalisation. If, however, a characterization is carried out with regional or temporal differentiation, the current flow and normalisation flow will differ if the environmental target is not also formulated on the basis of the characterised emissions. The characterised flow comprises only those substance flows for which the Eco Factors are determined via the characterization. In accordance with the highest eco-factor principle, Eco Factors are only ever assessed in relation to the most stringent target. If another target is more stringent for a given flow, normalisation may no longer contain this flow.

4.3.2   Determination of Weighting

National annual flows are generally used for weighting. Depending on the matter in hand, site-specific, regional, national, continental or global or seasonal or annual current and critical flows may be used for certain environmental problems. Flows are quantified either as individual substances or as environmental impacts in accordance with environmental targets and compatibly with normalisation. The weighting term is unitless, which means that current and critical flows must be stated in the same units. The weighting function is also quadratic for spatially and temporally differentiating Eco Factors. Current flows must always be determined in terms of the reduction target. The reference basis for current and critical flows should be identical. Current flow must be calculated in accordance with the target or the critical flow. Current flow is usually identical to the normalisation flow. Critical flows are generally based on politically binding objectives (which should in turn be based on scientific findings). These are primarily officially set protection targets (annual loads, ambient limit values). In the absence of direct official stipulations, political declarations of intent which are as binding as possible are used as the basis.

4.3.3   Eco-Factor Determination

Being the result of characterization, normalisation and weighting, Eco Factors represent the political and legislative assessment of the ecological significance of the pollutants. For example, emissions of various heavy metals into the air, soil and water are each assessed with a dedicated eco-factor which is (ideally) calculated from the respective current and critical flows. This usually means that a different eco-factor is obtained for emission of a single pollutant into water, air or soil. These differences reflect differing statutory requirements and current impacts.

4.3.4   Temporal Aspects of the Eco-Factor Dermination/ Time Horizons

Statutory stipulations, e.g. ambient limit values for air pollutants, generally do not contain an explicit time horizon, other than in transitional provisions. They apply from their effective date onwards. In the case of political stipulations, by contrast, specific targets may be defined for particular points in time. When determining an eco-factor, if several policy objectives with (very) different time horizons exist, then based on an appraisal of the current political situation, either one of the points in time should be selected or an intermediate point in time should be determined by interpolation.

Because the definition of target values is staggered in terms of both timing and subject, the "baselines"[1] as the starting points for minimization targets may differ greatly from one another depending on the environmental impact in question. It should be ensured in each case that the most recent definition is applied when determining the Eco Factors.

Note: The present report "Ecological Scarcity Method for the European Union" makes use of the targets and thus also the time horizons of the EU environmental authorities. It is in the nature of things that, for the authorities, these horizons each depend on the state of research and planning, implementation scenarios, urgency, current political relevance and votes as well as other factors. These scheduling variables differ for different environmental impacts, sometimes resulting in implementation targets with different time frames.

4.4 General Data Situation

Collecting the European eco-factor data set for the Ecological Scarcity Method primarily involves two major aspects:

- obtaining a statistical record of the actual state of annual loads and consumption figures and
- articulating the political will with regard to intended future conditions in the region under consideration.

---

1   i. e. years used as a basis for reference (frequently then set as 100%).

The EU procedure for collecting EU data is described in EEA 2009, which outlines the processes involved in requesting, checking and correcting the submissions from the individual countries. The EEA points out that there are gaps in the data. This is due to different levels of development of individual countries' recording systems and to new accessions by countries in which EU-compatible recording systems have yet to be established.

Similarly, not all environmental targets have yet been articulated in an operationally uniform way as desirable states for all major impacts. Quality targets sometimes compete with procedural targets while binding targets sometimes compete with non-binding ones. Some of the time horizons set for the objectives are also very different, since these definitions depend on the nature and duration of the reduction programmes, as well as on international coordination efforts and sometimes also on changing basic technological conditions and possibilities for countering the environmental impacts in question.

4.4.1 Reccording the Actual State

The data from participating countries are compiled and aggregated in various pro-grammes at the EU level. These data are collected at regular intervals, checked for completeness, subjected to plausibility checking during compilation and, if necessary, corrected with the assistance of the individual countries. These data are then also made publicly available.

4.4.2 Articulating Political Will

Political will is articulated and environmental policy targets thus defined at the EU level through the competent authorities. The European Commission has a number of Directorates-General which, working together with the associated authorities and institutes, devise targets in the fields of air pollution, water contamination, climate protection, resource scarcity, waste and others and implement them politically. Identical or similar measures are also implemented in the individual countries unless existing European Union targets are explicitly adopted.

4.4.3 Discussion of Procedure

At the project planning stage, the aim was to determine a set of Eco Factors for the entire EU-28 as a geographic region with its own environmental policy. Over the course of the project, it became clear that very large volumes of data are also available for individual EU countries arising from current environmental policy, social and national government activities. With certain limitations, these data make it possible to determine the respective Eco Factors at the national level too (cf. section 3.3.5).

On completion of the various analytical steps, the project partners were in agreement that, despite the mentioned limitations, it is certainly meaningful to determine the individual countries' Eco Factors and list them in the present study. This is justified by the better geographic resolution of environmental policy efforts and natural landscape circumstances, such as the very different availability of water in the individual countries, and by the consequently greater acceptance of the data by local companies.

5 Results Type of Impact

5.1 Emission to Air

5.1.1 Greenhouse Gases

Greenhouse gases are considered to have a major influence on climate warming. The Kyoto Protocol (United Nations Framework Convention on Climate Change, 1992) calls for reductions in the following greenhouse gases:

- carbon dioxide (CO_2)
- methane (CH_4)
- dinitrogen monoxide (N_2O)
- partially halogenated hydrofluorocarbons (HFC)
- perfluorinated hydrocarbons (PFC)
- sulfur hexafluoride (SF_6)

The Montreal Protocol on Substances that Deplete the Ozone Layer governs the reduction in CFC emissions and these substances, despite having a greenhouse effect, are thus not included in the Kyoto Protocol.

The Kyoto Protocol greenhouse gases (excluding CFCs) are abbreviated "GHG" below.

5.1.2 Characterization

The current publication of the "Intergovernmental Panel on Climate Change (IPCC)" (Forster, *et al.*, 2007) serves as the reference for the global warming potential (GWP) of various gases. The reference substance is carbon dioxide (CO_2). The potentials are adapted as required to new scientific findings and new substances are described. The values vary depending on the time period over which the effects are summed. GWP100 values (100-year integration time) are normally used (see Table 1) and are also used for characterization in the present report.

The European Environment Agency statement (EEA 2014) is used here. The European Commission objective is to achieve a reduction of 80–95% compared to 1990 emission levels by 2050 (European Commission EC 2011). In the present case a reduction in an amount of the minimum target value of 80% has been assumed in the first step, since it is at this level that the target worded in this way would first be achieved. If good progress is made, this objective may be further adjusted on subsequent revision of the data (cf. Ahbe *et al* 2014).

Table 1:   Global warming potential for various greenhouse gases[1]

Substance	Global warming potential (GWP) in CO2eq
CO_2	1
CH_4	25
N_2O	298
HFC	124–14,800
PFC	7,390–17,700
SF_6	22,800

Table 2:   Greenhouse gas emissions, EU-28

Reference year	EU-28 GHG emissions (excluding CFCs) in (million t CO_2eq)/y	Source
Baseline: 1990	5,626	EEA 2014
Current: 2012	4,544	EEA 2014
Target: 2050 (80%)	1,125	EC 2011

Applying the calculation specification, the following eco-factor is thus obtained, here shown purely by way of example:

$$EF_{CO2eq} = 1 \cdot \frac{EP}{4,544,000 * \frac{10^9 g}{a}} \cdot \left(\frac{4,544,000 * \frac{10^9 g}{a}}{1,125,000 * \frac{10^9 g}{a}} \right)^2 \cdot \frac{10^{12}}{a} = 0.00359 \frac{EP}{g}$$

Table 3:   EU-28 eco-factor for greenhouse gas emissions

Reference year	Eco-factor in EP, EU-28/g CO_2eq
2050	0.00359

5.1.3   Preliminary Remarks on Air Pollutants

There are essentially two competing perspectives on this impact category: on the one hand, the non-binding targets according to "Thematic Strategy on Air Pollution" ((TSAP), European Commission 2005) and, on the other, the more recent, binding, but less demanding reduction targets according to the Gothenburg Protocol (ECE/

---

1   (Forster, et al., 2007), p. 212

EB.AIR/114, 2012). In a written communication dated 10.7.2014, the EEA stated, in response to an enquiry, that the TSAP targets from 2005 should be considered outdated and that it was obvious that these targets could not be achieved within the originally planned periods. For this reason there should be a reorientation, which is described as follows:

> "The European Commission completed a review of the EU air legislation and TSAP at the end of 2013. It has proposed new legal EU targets for 2020 (based on the Gothenburg ceilings) and also for 2030 (representing a further step of moving towards the EU objective of ensuring "levels of air quality that do not give rise to significant negative impacts on, and risks to human health and the environment" (EEA 2014a).

The new, binding definition based on the Gothenburg Protocol is thus used below, since this definition has also been agreed by the EU with the individual countries with regard to implementation. The actual values used in these calculations are those from 2010, since they reflect the latest determined and established state. Since previous and future reduction efforts will in many cases result in lower annual loads, the eco-factor calculations will have to be adapted from time to time to take account of the actual, more recent emission loads.

5.1.4   NMVOC

NMVOC (non-methane volatile organic compounds):

In addition to their sometimes toxic effects, these are precursor substances for photo-oxidants and are harmful to human and plant health. According to the 31st Federal Pollution Control Ordinance (article 2, paragraph 11), a volatile organic compound is defined as "an organic compound which, at 293.15 kelvin, has a vapour impact of 0.01 kilopascals or more or, under its particular conditions of use, has a corresponding volatility. The creosote fraction which, at 293.15 kelvin, exceeds this vapour impact is considered to be a volatile organic compound." Methane is not subsumed in this category.

The intention is, by 2020, to reduce emissions by 28% in comparison with 2005 levels (ECE/EB.AIR/114, 2012).

Table 4:   NMVOC emissions, EU-28

Reference year	EU-28 NMVOC emissions in (kt VOC)/y	Source
Baseline: 2005	8,842	ECE/EB.AIR/114, 2012
Actual load: 2010	7,500	EEA (2013)
Target: 2020 (28%)	6,366	ECE/EB.AIR/114, 2012

Applying the calculation specification, the following eco-factor is thus obtained:

Table 5: EU-28 eco-factor for NMVOC emissions

Reference year	Eco-factor in EP, EU-28/g NMVOC
2020	0.1851

5.1.5 NO_x

NO_x (measured as NO_2) contributes not only to ozone formation potential (POCP) but also to eutrophication and acidification. The intention is, by 2020, to reduce emissions by 42% in comparison with 2005 levels (ECE/EB.AIR/114, 2012).

Table 6: NO_x emissions, EU-28

Reference year	EU-28 NO_x emissions in (kt NO_x)/y	Source
Baseline: 2005	11,355	ECE/EB.AIR/114, 2012
Actual load: 2010	9,000	EEA (2013)
Target: 2020 (42%)	6,585	ECE/EB.AIR/114, 2012

Applying the calculation specification, the following eco-factor is thus obtained:

Table 7: EU-28 eco-factor for NO_x emissions

Reference year	Eco-factor in EP, EU-28/g NO_x
2020	0.2076

5.1.6 SO_x

SO_x (measured as SO_2) are harmful to the respiratory organs and are a precursor substance for acidic precipitation which in turn damages plants and buildings. The intention is, by 2020, to reduce emissions by 59% in comparison with 2005 levels (ECE/EB.AIR/114, 2012).

Table 8: SO_2 emissions, EU-28

Reference year	EU-28 SO_2 emissions in (kt SO_2)/y	Source
Baseline: 2005	7,828	ECE/EB.AIR/114, 2012
Actual load: 2010	5,000	EEA (2013)
Target: 2020 (59%)	3,209	ECE/EB.AIR/114, 2012

Applying the calculation specification, the following eco-factor is thus obtained:

Table 9: EU-28 eco-factor for SO_2 emissions

Reference year	Eco-factor in EP, EU-28/g SO_2
2020	0.4855

5.1.7   PM2.5

Fine particulate matter consists of particles of very different origins: abrasion, geological and biological material, soot etc. Its effects range from health impairment via respiratory tract conditions to carcinogenicity. One distinguishing feature is maximum diameter, which is conventionally broken down into PM10 (particulate matter, max. 10 micrometres), PM2.5 and PM0.1 According to a communication from the German UBA (UBA, 2013c), in the absence of further information, a PM2.5 value can be estimated for approximation purposes by multiplying the PM10 value by a factor of 0.7.

The intention is, by 2020, to reduce emissions by 22% in comparison with 2005 levels (ECE/EB.AIR/114, 2012).

Table 10: PM2.5 emissions, EU-28

Reference year	EU-28 PM2.5 emissions in (kt PM2.5)/y	Source
Baseline: 2005	1,504	ECE/EB.AIR/114, 2012
Actual load: 2010	1,350	EEA (2013)
Target: 2020 (22%)	1,173	ECE/EB.AIR/114, 2012

Applying the calculation specification, the following eco-factor is thus obtained:

Table 11: EU-28 eco-factor for PM2.5 emissions

Reference year	Eco-factor in EP, EU-28/g PM2.5
2020	0.9812

5.1.8 NH_3

Ammonia (NH_3) is predominantly formed in agriculture, but also in industrial processes. NH_3 plays a major part in forming acidic precipitation and is harmful to the ecosystem not only by acidification and overfertilisation but also by the formation of secondary pollutants.

The intention is, by 2020, to reduce emissions by 6% in comparison with 2005 levels (ECE/EB.AIR/114, 2012).

Table 12: NH₃ emissions, EU-28

Reference year	EU-28 NH₃ emissions in (kt NH₃)/y	Source
Baseline: 2005	3,813	ECE/EB.AIR/114, 2012
Actual load: 2010	3,500	EEA (2013)
Target: 2020 (6%)	3,584	ECE/EB.AIR/114, 2012

Applying the calculation specification, the following eco-factor is thus obtained:

Table 13: EU-28 eco-factor for NH₃ emissions

Reference year	Eco-factor in EP, EU-28/g NH₃
2020	0.2725

5.2 Emissions Surface Water

At EU-28 level, various recording systems have been used over an extended period. The EEA recommends using the current system, the "European Pollutant Release and Transfer Register" (E-PRTR) (EEA 2014b). This register replaced the preceding system EPER[2] in 2009 and was designed to provide the public and relevant bodies with improved access to environmental data. The EEA simultaneously pointed out, however, that this register does not record all emission sources. The emphasis here is on relatively large industrial facilities that exceed a certain specified annual load threshold as "point sources". The E-PRTR system consequently does not include "diffuse sources", which also include those point dischargers which are below the reporting threshold. The EEA has carried out further investigations to identify the influence of "diffuse sources" and published the findings as a report (Deltares 2013).

The Deltares report investigated the proportion of diffuse sources in various discharge routes above and beyond the data already reported and recorded by the E-PRTR database (emissions from large point sources). Due a lack of data, the Deltares report explicitly provides no information about diffuse industrial emissions (i.e. those below the E-PRTR recording threshold), which nevertheless play a significant role in water pollution.

---

2 European Pollutant Emission Register

In order to make good this shortcoming, German conditions were used as the basis for the following extrapolation: a factor corresponding to the ratio of the German values for emissions to surface water coordinated with the UBA to the German data from the Deltares report and the E-PRTR database was calculated for each emission (each water pollutant). These factors were then multiplied with those from the Deltares report and the E-PRTR database values. The proportion of diffuse industrial emissions derived from the better coordinated German data was thus applied in an identical ratio by extrapolation to EU-28. Extrapolation is thus carried out for each pollutant using the following scheme:

Table 14: Extrapolation scheme for pollutant discharge to surface water, EU-28

Proportion from	Calculation	
E-PRTR (2012)		Proportion of large point sources above threshold value
Deltares (2013)	+	Proportion of diffuse sources and smaller point sources excluding industrial sources
Extrapolation factor, determined from German data	+	Proportion of diffuse and smaller point sources from industry
Actual total, EU-26 (excl. HR & MT)	=	Total EU-26 excluding Croatia and Malta
Extrapolation factor for Croatia and Malta*	1.009	
Total actual load EU-28	=	Total EU-28 (incl. HR & MT)
Reduction target for this pollutant, determined from German data	Reduction ratio* = Fk/F (from D)	
Total target load (Fk)	=	Actual load EU-28*(Fk/F)$_D$

*) pro rata to population (Eurostat 2014)

Target values: in response to a question, the EEA stated in a written communication of 2.10.2014 that no reduction targets for emissions to surface water have yet been defined as annual loads at EU-28 level (EEA 2014d). Various more local reduction programmes are being carried out, for example by the states bordering the North Atlantic or the Baltic Sea. The target values were thus selected in such a way that the ratios of actual and target values per impact correspond to those of the respective German data coordinated with the Federal Environment Agency.

The E-PRTR and Deltares study tables of values do not include any data for Croatia (HR) and Malta (MT). The data from the EU-26 format have therefore been extra-polated pro rata by population to EU-28 level.

This approach involves considerable inaccuracy in relation to these two individual countries. It is, however, better in the present case to make use of such an extrapolation than simply to treat the unknown quantities as non-existent and omit them. The investigations and discussions with the EEA have revealed that there is indeed still a need for further research and clarification in relation to water pollution. This need primarily relates firstly to recording current pollutant loads as completely as possible and secondly to defining objectives in order to specify desirable conditions for the future. The current state of the environment including all major pollutant loads, and the desired state of the environment, which can then be appropriately used for defining targets and as a guidance benchmark for industry, both have to be defined as clearly as possible to provide a basis for assessing industrial environmental impacts. Carrying out further research in this direction would have gone beyond the scope and time horizon of the present paper. The values were therefore extrapolated as described, subject to the described limitations.

Note regarding extrapolation by population size: Where country data are unavailable, a range of criteria may be used for extrapolating environmental impacts: population size, GDP, per capita GDP, energy generation or waste treatment parameters and many others, taken alone or in combination. Since plant technology and environmental regulations also vary considerably, it is largely impossible to state a single, generally applicable extrapolation formula. For reasons of simplicity, the ratio of populations was accordingly selected as the extrapolation criterion for the present paper.

5.2.1   Nitrogen (as N)

Surface waters have variable sensitivity towards nitrogen compounds and two different effects are of significance: acidification and eutrophication.

Lakes which are naturally nitrogen-limited either year-round or temporarily are severely affected by nitrogen inputs. One major problem that extends beyond national borders is elevated nitrogen inputs to the oceans and their associated eutrophication. Nitrogen primarily enters the oceans via rivers, but also via the atmosphere, and, once in the oceans, it is the growth-limiting and hence the decisive factor for nutrient effects (UBA, 2009), (UBA, 2011). The sources provide the following data:

Table 15: Nitrogen emissions to surface water, EU-28

Proportion from	EU-28: N emissions (kt N)/y	Source
E-PRTR (2012)	384.8	EEA (2014a)
Deltares (2013)	3,388	Deltares (2013)
Extrapolation factor = 1.68	2,557	Ahbe *et al*. (2014)
Actual total, EU-26 (excl. HR & MT)	6,330	-
Actual total, EU-28 (incl. HR & MT)	6,387	-
Target (reduction ratio D = 0.913)	5,831	Ahbe *et al*. (2014)

Applying the calculation specification, the following eco-factor is thus obtained:

Table 16: EU-28 eco-factor for nitrogen emissions

Reference year	Eco-factor in EP, EU-28/g N
2014	0.1879

5.2.2   Phosphorus (as P)

Elevated nutrient inputs (nitrogen and phosphorus inputs) into rivers, lakes, coastal waters and oceans generally result in eutrophication of these bodies of water (cf. 5.2.1 Nitrogen). The growth-limiting nutrient is usually phosphorus (UBA, 2009), (UBA, 2011).

The sources provide the following data:

Table 17: Phosphorus emissions to surface water, EU-28

Proportion from	EU-28: P emissions (kt P)/y	Source
E-PRTR (2012)	40.2	EEA (2014a)
Deltares (2013)	131.9	Deltares (2013)
Extrapolation factor = 1.89	153.17	Ahbe *et al.* (2014)
Actual total, EU-26 (excl. HR & MT)	325.3	-
Actual total, EU-28 (incl. HR & MT)	327.4	-
Target (reduction ratio D = 0.397)	130.1	Ahbe *et al.* (2014)

Applying the calculation specification, the following eco-factor is thus obtained:

Table 18: EU-28 eco-factor for phosphorus emissions

Reference year	Eco-factor in EP, EU-28/g P
2014	19.34

5.2.3   Nickel

"Heavy metals, being chemical elements, are not degradable in the environment. They only become a hazard for humans and the environment at elevated concentrations and if they can be absorbed by living organisms. Environmental concentrations are usually so low that no acute toxic effects occur. Long-term, chronic toxic effects, on the other hand, are to be anticipated if individual heavy metals are able to enter the food chain and accumulate in living organisms." (Hesse State Agency for the Environment and Geology (DE), 2013).

"When making an ecological assessment, the focus should primarily be on the phyto-toxicity of the heavy metals copper, chromium, nickel and zinc, but on the human or animal toxicity of cadmium and lead." (State Agency for the Environment, Nature Conservation and Geology (DE), 2002).

Nickel is moreover one of the 33 priority substances under EU law (Directive 2008/105/EC of the European Parliament and of the Council on environmental quality standards in the field of water policy, amending Directive 2000/60/EC).

The sources provide the following data:

Table 19: Nickel emissions to surface water, EU-28

Proportion from	EU-28: Ni emissions (t Ni)/y	Source
E-PRTR (2012)	263.0	EEA (2014a)
Deltares (2013)	472.9	Deltares (2013)
Extrapolation factor = 4.68	2,708	Ahbe et al. (2014)
Actual total, EU-26 (excl. HR & MT)	3,444	-
Actual total, EU-28 (incl. HR & MT)	3,472	-
Target (reduction ratio D = 0.472)	1,638	Ahbe et al. (2014)

Applying the calculation specification, the following eco-factor is thus obtained:

Table 20: EU-28 eco-factor for nickel emissions

Reference year	Eco-factor in UBP, EU-28/g Ni
2014	1,293

5.2.4   Zinc

Zinc discharges to surface water contribute to harming plant growth. The sources provide the following data:

Table 21: Zinc emissions to surface water, EU-28

Proportion from	EU-28: Zn emissions (t Zn)/y	Source
E-PRTR (2012)	1,796	EEA (2014a)
Deltares (2013)	2,619	Deltares (2013)
Extrapolation factor = 4.38	14,923	Ahbe *et al.* (2014)
Actual total, EU-26 (excl. HR & MT)	19,338	-
Actual total, EU-28 (incl. HR & MT)	19,506	-
Target (reduction ratio D = 0.64)	12,491	Ahbe *et al.* (2014)

Applying the calculation specification, the following eco-factor is thus obtained:

Table 22: EU-28 eco-factor for zinc emissions

Reference year	Eco-factor in EP, EU-28/g Zn
2014	125.0

5.2.5   COD

DIN 38 409, part 41 (Deutsches Institut für Normung e.V., 1980) defines chemical oxygen demand (COD) as follows:

"The chemical oxygen demand (COD) of a volume of water is taken to be the volume-related mass of oxygen which is equivalent to the mass of potassium dichromate which, under the working conditions of the method, reacts with the oxidisable substances present in the water." COD indicates the quantity of oxygen which is required to oxidise the organic compounds present in the water and is thus a measure of the concentration of organic substances in water.

All organic substances cause water pollution by oxygen consumption and thus limit the habitat for the fauna. Many substances may additionally have specific toxic effects which would have to be separately recorded (Frischknecht, Steiner, & Jungbluth, 2009).

The sources provide the following data:

Table 23: COD emissions to surface water, EU-28

Proportion from	EU-28: COD emissions (kt COD)/y	Source
E-PRTR (2012)	1,669	EEA (2014a)
Deltares (2013)	1,800	Deltares (2013)
Extrapolation factor = 1.18	624.4	Ahbe *et al.* (2014)
Actual total, EU-26 (excl. HR & MT)	4,093	-
Actual total, EU-28 (incl. HR & MT)	4,128	-
Target (reduction ratio D = 0.54)	2,226	Ahbe *et al.* (2014)

Applying the calculation specification, the following eco-factor is thus obtained:

Table 24: EU-28 eco-factor for COD emissions

Reference year	Eco-factor in EP, EU-28/g COD
2014	0.833

5.2.6   Lead

Humans, animals and plants can be harmed by emissions of lead. Lead is capable of accumulating in the food chain and as a result becoming more concentrated in the organism consuming it. The sources provide the following data:

Table 25: Lead emissions to surface water, EU-28

Proportion from	EU-28: Pb emissions (t Pb)/y	Source
E-PRTR (2012)	175.0	EEA (2014a)
Deltares (2013)	452.2	Deltares (2013)
Extrapolation factor = 3.9	1,819	Ahbe *et al.* (2014)
Actual total, EU-26 (excl. HR & MT)	2,446	-
Actual total, EU-28 (incl. HR & MT)	2,469	-
Target (reduction ratio D = 0.25)	617.1	Ahbe *et al.* (2014)

Applying the calculation specification, the following eco-factor is thus obtained:

Table 26: EU-28 eco-factor for lead emissions

Reference year	Eco-factor in EP, EU-28/g Pb
2014	6,483

5.2.7 Cadmium

Cadmium has a toxic effect in humans and animals and can bioaccumulate, disrupting metabolic processes and possibly being carcinogenic. The sources provide the following data:

Table 27: Cadmium emissions to surface water, EU-28

Proportion from	EU-28: Cd emissions (t Cd)/y	Source
E-PRTR (2012)	26.2*	EEA (2014a)
Deltares (2013)	52.56*	Deltares (2013)
Extrapolation factor = 1.0	-	Ahbe *et al.* (2014)
Actual total, EU-26 (excl. HR & MT)	78.76	-
Actual total, EU-28 (incl. HR & MT)	79.55	-
Target (reduction ratio D = 0.25)	19.91	Ahbe *et al.* (2014)

*) numbers rounded

Applying the calculation specification, the following eco-factor is thus obtained:

Table 28: EU-28 eco-factor for cadmium emissions:

Reference year	Eco-factor in EP, EU-28/g Cd
2014	200,705

5.2.8 Copper

Even relatively low concentrations of copper emissions can have a disruptive and harmful effect on aquatic systems.

The sources provide the following data:

Table 29:   Copper emissions to surface water, EU-28

Proportion from	EU-28: Cu emissions (t Cu)/y	Source
E-PRTR (2012)	418.0	EEA (2014a)
Deltares (2013)	642.02	Deltares (2013)
Extrapolation factor = 3.12	2,252	Ahbe *et al.* (2014)
Actual total, EU-26 (excl. HR & MT)	3,312	-
Actual total, EU-28 (incl. HR & MT)	3,341	-
Target (reduction ratio D = 0.765)	2,556	Ahbe *et al.* (2014)

Applying the calculation specification, the following eco-factor is thus obtained:

Table 30:   EU-28 eco-factor for copper emissions

Reference year	Eco-factor in EP, EU-28/g Cu
2014	511.2

5.2.9   EPA-PAH16

PAH are usually formed during the combustion of hydrocarbons and have a toxic and sometimes carcinogenic effect. PAH are various forms of fused benzene rings and several hundred of these compounds are known. Depending on purpose, various cumulative parameters, such as in this case EPA-PAH16, may be used for the substantially occurring compounds.

There are no statistical surveys of EPA-PAH16 emissions at EU-28 level, nor have any targets been set which can be used in this context. Omitting this environmental impact for this reason would considerably magnify the relative error in the assessments. Therefore, by way of a provisional solution, German conditions have been extrapolated pro rata to population to European conditions. It should be noted that, here too, this can at best be an approximation, constituting a first step in establishing the ESM assessment method with European data. The data must, however, be stated more accurately when the collected data and stated targets are updated in the future.

Table 31: EPA-PAH16 emissions to surface water, EU-28

	EPA-PAH16 in t/y	Population	Source
Germany, actual load 2005	19.16		Ahbe *et al.* (2014)
Germany, target load	4.41		Ahbe *et al.* (2014)
Population, EU-28		503 million	Eurostat, population 2010
Population, DE		81.8 million	Eurostat, population 2010
Extrapolation, actual load EU-28	117.82		
Extrapolation, target load EU-28	27.14		

Applying the calculation specification, the following eco-factor is thus obtained:

Table 32: EU-28 eco-factor for EPA-PAH16 emissions

Reference year	Eco-factor in EP, EU-28/g EPA-PAH16
2014	160,099

5.3 Consumption of Resources

5.3.1 Freshwater Consumption

Freshwater consumption is defined (OECD, 2013) as any extraction of freshwater for production or consumption processes. Water used for power generation in hydro-electric power stations is not included in this definition.

The EEA provides information at relatively long intervals about absolute water extraction volumes by EU countries and about the countries' water exploitation index (WEI), which characterises the ratio of extraction volume to water supply, i.e. to the volume available to a country over the long term. The present paper adopts the OECD criterion necessary for defining "critical consumption" (Fk) and defines a WEI of 20% as the limit of tolerability ((OECD 2013), (Ahbe *et al.* 2014)).

Table 33: OECD definition of water scarcity

OECD definition of water scarcity	WEI	Source
Moderate	10–20%	OECD (2008)
Medium	20–40%	OECD (2008)
High	over 40%	OECD (2008)
Critical extraction criterion (moderate to medium)	20%	OECD (2013)

The following values are obtained for the corresponding volumes:

Table 34: Water scarcity, EU-28

Water scarcity, EU-28	Billion m³/y	Source
EU-28 water extraction	270.8	EEA 2012, 2010
EU-28 water supply	3,119	EEA 2012
Critical extraction volume, EU-28	623.8	EEA 2012, 2010

Applying the calculation specification, the following eco-factor is thus obtained:

Table 35: EU-28 eco-factor for freshwater consumption

Reference year	Eco-factor in EP, EU-28/m³
2010	0.6959

5.3.2   Primary and Renewable Energy Consumption

The assumption in ESM assessments is that primary energy consumption should be reduced, while the proportion of "renewable energy consumption" in primary energy consumption should increase. This means that, from the standpoint of the environmental impact assessment user, two Eco Factors must be stated with regard to energy scarcity which take account firstly of the described targets and secondly of the fact that the scarcity situation of the resource primary energy becomes less severe as renewable energies gradually replace the consumption of non-renewable primary energy.

An eco-factor is therefore stated both for "non-renewable primary energy consumption" and for "renewable energy consumption" (cf. also Ahbe et al. 2014).

The corresponding actual energy consumption values are obtained from stated Eurostat values for primary energy and renewable energies.

Table 36: Actual energy consumption, EU-28

Actual energy consumption in EU-28 for 2012	PJ/y	Source
Actual primary energy consumption	70,460	Eurostat 2014a
Actual renewable energy consumption	7,429	Eurostat 2014b
Actual non-renewable primary energy consumption	63,031	Line 1 minus line 2

Directive 2012/27/EU states a target value for total primary energy consumption in 2020. Directive 2009/28/EC sets country-specific targets for the proportion of final energy consumption to be accounted for by renewable sources. These proportions were applied to the calculated target values for total primary energy consumption for 2020 and multiplied by the ratio of EU-28 primary energy consumption to final energy consumption in order to obtain the desired value for the proportion of primary energy consumption accounted for by renewable sources in 2020.

This calculation thus initially assumes that the ratio of primary energy to final energy will remain constant until 2020. Strictly speaking, this is not the case because the difference between the two types of energy largely consists of power station losses, the minimisation of which is currently being vigorously addressed. The timescale for upgrading power station technologies to distinctly better efficiency levels is, however, rather long. The change in the ratio of primary to final energy consumption which is to be anticipated by 2020 is thus probably not so great as to make the simplification selected here for lack of better data no longer acceptable.

Non-renewable primary energy is obtained from the difference between total primary energy and renewable sources.

Table 37: Target energy consumption, EU-28

Target energy consumption in EU-28 for 2020	PJ/y	Source
Target primary energy consumption	61,713	Directive 2012/27/EU
Target renewable energy consumption	8,422	Directive 2009/28/EC
Target non-renewable primary energy consump.	53,291	Line 1 minus line 2

Applying the calculation specification, an eco-factor for consumption of "non-renewable energy sources" may be calculated (cf. Ahbe *et al.* 2014). Amounting to 8422 PJ/y for 2020, this constitutes 13.6% of what will then be the definitive target primary energy consumption of 61,713 PJ/y. In order to determine the eco-factor for "renewable energy consumption", the value of the eco-factor for consumption of "non-renewable primary energy" is reduced by this value of 13.6%, resulting in a value of 0.01917 EP/MJeq.

Table 38: EU-28 Eco Factors for energy consumption

Reference year 2020	Eco-factor in EP, EU-28/MJeq
Renewable energy consumption	0.01917
Non-renewable primary energy consumption	0.02219

First note: The time horizon for the primary energy consumption objectives is 2050 for the German data set approved by the UBA and 2020 for the remainder of Europe. No targets extending beyond that point have (yet) been defined at a European level. The target of reducing European greenhouse gas emissions by 80–95% by 2050, does not allow any direct conclusions to be drawn regarding a reduction target for primary energy over the same period, since there are various sources of greenhouse gases and CO_2 storage technology cannot be ruled out for this time horizon. While the EU is indeed responsible for defining objectives in the field of greenhouse gas emissions, energy policy is a matter for the individual countries, which explains the absence of more far-reaching, binding objectives in European energy policy.

Second note: The values listed above solely reflect the energy scarcity situation. When calculating the entire environmental impact caused by consuming non-renewable (in particular fossil) energy, it is additionally necessary to take account of the other environmental impacts such as pollutant emissions by applying the corresponding Eco Factors stated above (cf. Ahbe *et al.* 2014).

5.4 Waste Generation

5.4.1   Non-Hazardous and Hazardous Waste

In addition to the intended reduction in waste volumes, the European Commission is also planning to move towards resource cycle management, in which waste becomes a raw material again, so changing the meaning of the word "waste". The target is to utilise the greatest possible proportion of the generated waste streams by means of material recycling or energy recovery[3].

Hazardous and non-hazardous waste is treated separately from a waste management standpoint and separate Eco Factors have accordingly been determined.

A distinction is drawn below between

- non-hazardous waste and
- hazardous waste.

---

3   cf. "Being wise with waste: the EU's approach to waste management", European Union 2010

The volume of non-hazardous waste is determined by subtracting hazardous waste, which is considered separately, from the total volume of waste generated. Waste from mining, from the extraction of rock and ore and from the construction sector are also subtracted because they are largely mineral in nature and virtually unreactive.

With regard to the objectives for both non-hazardous and hazardous waste, it is assumed, as in the German Federal Environment Agency definition (cf. Ahbe *et al.* 2014), that the present day waste volumes described above should be considered more or less critical and should therefore not be allowed to rise further in future. In other words, the current flow and critical flow are identical and this applies to both classes of waste. In line with a written statement from the EEA in July 2014, this assessment is also applied at EU-28 level (EEA 2014c).

The following waste volume values are obtained:

Table 39:   Waste generation, EU-28

Waste generation, EU-28	Mt/y	Source
Non-hazardous waste, 2010	893.5	Eurostat 2014c
Hazardous waste, 2010	94.46	Eurostat 2014c

Table 40:   Critical waste volume, EU-28

Critical waste volume, EU-28	Mt/y	Source/details
Non-hazardous waste	893.5	EEA, July 2014
Hazardous waste	94.46	EEA, July 2014

Applying the calculation specification, the following Eco Factors for waste are obtained:

Table 41:   Eco Factors for waste, EU-28

	Eco-factor in EP, EU-28/g
Non-hazardous waste	0.00112
Hazardous waste	0.01059

5.5 Derived Data Sets for Individual EU Countries

5.5.1   References to Calculation in the Datasheets

General

■ ISO 3166 alpha-2 country codes are used.

■ The data sets have not been coordinated in this form with the authorities of the individual countries (cf. item 3.3.5).

Calculation of individual values

■ The EU target is to reduce greenhouse gas emissions by 80–95% from their 1990 levels by 2050. The 80% target has been selected for the individual countries too. Should it prove possible to "exceed" the target reduction, the target value can be further adjusted on subsequent review.

■ The Deltares report data originate from 2005–2010, those for E-PRTR from 2012.

■ The conversion factor for TOC into COD is 3, a value which was obtained from the E-PRTR database (European Environment Agency 2014a).

■ EPA-PAH16: the Deltares report lists only anthracene and fluoranthene. Since there is no suitable method with which total PAH16 emissions can be estimated on the basis of these two substances, the PAH16 values for the individual countries have been provisionally extrapolated pro rata to population from the German values.

■ Freshwater: the OECD stress limit of 20%, indicating what is considered to be the critical flow, has been applied to all countries.

Specific features of the values for individual countries

■ Extrapolation of emissions to water for Malta and Croatia:
  – Croatia: is not yet included in the Deltares report and as yet has no entries in the E-PRTR database. Values for emissions to water in Croatia are therefore calculated pro rata to population (2014) on the basis of determined total EU-27 emissions (excluding Malta) (source for population sizes: Eurostat).
  – Malta: since the Deltares report lists too few values for Malta and the E-PRTR database likewise states no emission values for the pollutants (too few large industrial plants with an obligation to report), there is no reasonable basis for calculating emissions to water. Values were therefore extrapolated pro rata to population in the same way as for Croatia.

■ Waste in Estonia and Bulgaria: since the single value for hazardous waste in Estonia is very high, the EU average pro rata to population was used instead, thereby evening out the data on a plausibility basis. No further investigations were carried out into the causes of the data deviations. Since the value for mining waste in the statistics for Bulgaria is very high, a very low value for non-hazardous waste is obtained. The EU average pro rata to population for mining waste was therefore used instead. This, however, in turn resulted in a relatively high value for non-hazardous waste. No further investigations were carried out into the causes of the data deviations.

■ Renewable energy targets: on the basis of existing data, the calculated target value has already been exceeded in some countries (Czech Republic, Estonia, Croatia, Italy, Latvia, Lithuania, Hungary, Austria, Poland, Portugal, Romania, Slovakia, Finland and Sweden). This has been noted as a positive development and has not been further investigated.

■ Freshwater consumption in Cyprus: the relatively high value for the eco-factor is obtained because Cyprus' freshwater resources, with water consumption index of 63%, are under severe stress.

■ Freshwater consumption in Croatia: the average values for Italy, Slovenia and Greece as neighbouring countries were applied pro rata to population to Croatia. The reference countries were selected to provide similar climatic conditions.

■ Freshwater consumption in the UK: since the source used only gives details for England and Wales, a value pro rata to population was extrapolated to the UK.

6 Eco Factors for EU-28 and Member States

6.1 EU-28 (Regarded as one Environmentally Decision-Making Unit)

Environmental impact	Current flow	Critical flow	Eco-factor: EF-EU28-2014 (EP)	
Emissions to air:				
GHG CO_2eq [Mt/y]	4'544	1'125	0.00359	/g
NMVOC [kt/y]	7'500	6'366	0.1851	/g
NO_x [kt/y]	9'000	6'585	0.2076	/g
SO_2 [kt/y]	5'000	3'209	0.4855	/g
PM2.5 [kt/y]	1'350	1'173	0.9812	/g
NH_3 [kt/y]	3'500	3'584	0.2725	/g
Emissions to surface water:				
Nitrogen (as N) [kt/y]	6'387	5'831	0.1879	/g
Phosphorus (as P) [kt/y]	327.4	130.1	19.34	/g
Nickel [t/y]	3'472	1'638	1'293	/g
Zinc [t/y]	19'506	12'491	125.0	/g
COD [kt/y]	4'128	2'226	0.833	/g
Lead [t/y]	2'469	617.1	6'483	/g
Cadmium [t/y]	79.55	19.91	200'705	/g
Copper [t/y]	3'341	2'556	511.2	/g
EPA-PAH 16 [t/y]	117.9	27.14	160'099	/g
Resources				
Freshwater consumption [million m^3/y]	270.8	623.8	0.6959	/m^3
Energy efficiency/scarcity:				
Primary energy carriers [PJ/y]	70'460	61'713		
Consump. of renewable energy [PJ/y]	7'429	8'422	0.01917	/MJ
Non-renewable prim. energy carr. [PJ/y]	63'031	53'291	0.02219	/MJ
Waste				
Waste, non-hazardous [Mt/y]	893.5	893.5	0.00112	/g
Waste, hazardous [Mt/y]	94.46	94.46	0.01059	/g

6.2 Data sets of the EU Member States

The Eco Factors shown in the following tables have been calculated on the basis of publicly available data from EU databases or have been extrapolated in the manner stated in each case. The current values and targets shown therefore do not necessarily correspond to, and may thus vary from, the current environmental policy efforts of the individual country in question. The Eco Factors apply in each case to the stated country and, due to the differences in underlying data and total volumes, cannot be numerically compared with those for other countries. Further information can be found in the "explanations regarding data collection" section.

Legend of data tables (with the exception of Germany):

CuF: Current Flow, CrF: Critical Flow.

1) CuF: 2012 values according to the European Environment Agency (EEA)
   CrF: The EU has set itself the target of cutting greenhouse gas emissions by 80–95% compared to a 1990 baseline by 2050. A reduction target of 80% has been assumed for these data sets.

2) Gothenburg Protocol (ECE/EB.AIR/114, 2013)
   CuF: 2005 values
   CrF: Country-specific reduction targets for 2020 compared to 2005 values.

3) CuF: In order to fill gaps in the recording of industrial emissions, extrapolation has been performed from Deltares report data (EEA 2013) and the E-PRTR database for the countries on the basis of the values approved by the German UBA (Deltares report data originate from various years between 2005 and 2010. Data from the E-PRTR database relate to 2012).
   CrF: Since no European reduction targets have been defined, the target values were selected such that the ratios between actual and target values correspond to the German data approved by the UBA.

4) CuF/CrF: Values were calculated pro rata to population from the German data for 2010 approved by the UBA.

5) CuF: EEA (2002)
   CrF: A water consumption index (ratio of annual freshwater extraction to annual, long-term availability of freshwater) of 20% is considered to be the freshwater resource stress limit (OECD Guideline).

6) CuF: Eurostat "ten00086" and "ten00081". Values for 2012
   CrF: Primary energy carriers: The target value for the total consumption of EU28 in 2020 according to Directive 2009/27/EU has been broken down corresponding to the existing country percentages of the total of 2012. Consumption of renewable

energy: In Directive 2009/28/EC country-specific targets for the shares of renewable energy of final energy consumption have been fixed. These shares have been applied to the calculated targets for the total primary energy consumption of 2020 and were multiplied with the EU28 ratio of primary energy consumption and final energy consumption to obtain the striven percentage of renewable energy of the primary energy consumption. Not renewable primary energy carriers: Difference of primary energy carriers and consumption of renewable energy

7) CuF: Eurostat "env_wasgen". Values of 2010. For the evaluation of the amount of not harzardous waste the shares of hazardous waste, mining waste, exploitation of stones and ores as well as waste of building industry have been substracted from the total appearance of waste.

CrF: Targets for both kinds of waste (hazardous and not hazardous): The current quantities (without mineral shares) are regarded as critical and should not be exceeded (that means CuF=CrF) according to information of EEA of march 2014

6.2.1   Austria

Environmental impact	Current flow	Critical flow	Eco-factor: EF-AT-2014 (EP)
Emissions to air:			
GHG CO$_2$eq [Mt/y] 1)	80.10	15.62	0.3283 /g
NMVOC [kt/y] 2)	162	128	9.888 /g
NO$_x$ [kt/y] 2)	231	145.5	10.91 /g
SO$_2$ [kt/y] 2)	27	20	67.50 /g
PM2.5 [kt/y] 2)	22	17.60	71.02 /g
NH$_3$ [kt/y] 2)	63	62.40	16.18 /g
Emissions to surface water:			
Nitrogen (as N) [kt/y] 3)	120.7	110.2	9.943 /g
Phosphorus (as P) [kt/y] 3)	10.58	4.204	598.6 /g
Nickel [t/y] 3)	45.90	21.66	97'833 /g
Zinc [t/y] 3)	669.9	429.0	3'640 /g
COD [kt/y] 3)	84.55	45.59	40.67 /g
Lead [t/y] 3)	39.59	9.896	404'262 /g
Cadmium [t/y] 3)	1.313	0.3286	12'161'250 /g
Copper [t/y] 3)	52.54	40.20	32'507 /g
EPA-PAH 16 [t/y] 4)	1.997	0.4599	9'442'830 /g
Resources			
Freshwater consumption [million m^3/y] 5)	3.668	14.67	17.04 /m^3
Energy efficiency/scarcity: 6)			
Primary energy carriers [PJ/y]	1'409	1'234	
Consump. of renewable energy [PJ/y]	402.9 *)	275.2	0.8509 /MJ
Non-renewable prim. energy carr. [PJ/y]	1'006	958.5	1.095 /MJ
Waste			
Waste, non-hazardous [Mt/y] 7)	24.13	24.13	0.04144 /g
Waste, hazardous [Mt/y] 7)	1.473	1.473	0.6789 /g

*) According to the existing data-sources the calculated target value has been already exceeded.

6.2.2   *Belgium*

Environmental impact	Current flow	Critical flow	Eco-factor: EF-BE-2014 (EP)
Emissions to air:			
GHG CO$_2$eq [Mt/y] 1)	116.5	28.60	0.1424 /g
NMVOC [kt/y] 2)	143	113.0	11.20 /g
NO$_x$ [kt/y] 2)	291	171.7	9.872 /g
SO$_2$ [kt/y] 2)	145	82.65	21.23 /g
PM2.5 [kt/y] 2)	24	19.20	65.10 /g
NH$_3$ [kt/y] 2)	71	69.58	14.67 /g
Emissions to surface water:			
Nitrogen (as N) [kt/y] 3)	96.77	88.33	12.40 /g
Phosphorus (as P) [kt/y] 3)	3.620	1.439	1'749 /g
Nickel [t/y] 3)	62.30	29.40	72'081 /g
Zinc [t/y] 3)	641.2	410.6	3'803 /g
COD [kt/y] 3)	60.14	32.43	57.18 /g
Lead [t/y] 3)	23.71	5.926	675'058 /g
Cadmium [t/y] 3)	1.071	0.2680	14'906'865 /g
Copper [t/y] 3)	46.40	35.50	36'811 /g
EPA-PAH 16 [t/y] 4)	2.593	0.5969	7'275'261 /g
Resources			
Freshwater consumption [million m^3/y] 5)	6.217	3.886	411.8 /m^3
Energy efficiency/scarcity: 6)			
Primary energy carriers [PJ/y]	2'358	2'065	
Consump. of renewable energy [PJ/y]	117.9	176.1	0.5745 /MJ
Non-renewable prim. energy carr. [PJ/y]	2'240	1'889	0.6281 /MJ
Waste			
Waste, non-hazardous [Mt/y] 7)	38.58	38.58	0.02592 /g
Waste, hazardous [Mt/y] 7)	1.992	1.992	0.5020 /g

6.2.3   *Bulgaria*

Environmental impact	Current flow	Critical flow	Eco-factor: EF-BG-2014 (EP)
Emissions to air:			
GHG CO_2eq [Mt/y] 1)	61	21.82	0.1281 /g
NMVOC [kt/y] 2)	158	124.8	10.14 /g
NO_x [kt/y] 2)	154	90.86	18.65 /g
SO_2 [kt/y] 2)	777	170.9	26.59 /g
PM2.5 [kt/y] 2)	44	35.20	35.51 /g
NH_3 [kt/y] 2)	60	58.20	17.71 /g
Emissions to surface water:			
Nitrogen (as N) [kt/y] 3)	92.47	84.41	12.98 /g
Phosphorus (as P) [kt/y] 3)	4.081	1.622	1'552 /g
Nickel [t/y] 3)	42.36	19.99	106'021 /g
Zinc [t/y] 3)	232.2	148.7	10'502 /g
COD [kt/y] 3)	39.03	21.05	88 /g
Lead [t/y] 3)	38.38	9.593	417'046 /g
Cadmium [t/y] 3)	1.335	0.3340	11'961'663 /g
Copper [t/y] 3)	119.6	91.49	14'284 /g
EPA-PAH 16 [t/y] 4)	1.775	0.4087	10'625'927 /g
Resources			
Freshwater consumption [million m^3/y] 5)	5.934	19.78	15.17 /m^3
Energy efficiency/scarcity: 6)			
Primary energy carriers [PJ/y]	763.4	668.4	
Consump. of renewable energy [PJ/y]	68.58	70.16	1.738 /MJ
Non-renewable prim. energy carr. [PJ/y]	694.8	598.2	1.941 /MJ
Waste			
Waste, non-hazardous [Mt/y] 7)	112.4 *)	112.4 *)	0,008894 /g
Waste, hazardous [Mt/y] 7)	13.54	13.54	0.07384 /g

*) The Eurostat values for waste of mining and exploitation of stones and ores have been unreasonably high. Thus default values for these two positions have been calculated that meet the EU average quantity per person.

6.2.4 Croatia

Environmental impact	Current flow	Critical flow	Eco-factor: EF-HR-2014 (EP)
Emissions to air:			
GHG CO_2eq [Mt/y] 1)	26.40	6.380	0.6486 /g
NMVOC [kt/y] 2)	101	66.66	22.73 /g
NO_x [kt/y] 2)	81	55.89	25.93 /g
SO_2 [kt/y] 2)	63	28.35	78.39 /g
PM2.5 [kt/y] 2)	13	10.66	114.4 /g
NH_3 [kt/y] 2)	40	39.60	25.51 /g
Emissions to surface water:			
Nitrogen (as N) [kt/y] 3)	53.47 *)	48.81	22.44 /g
Phosphorus (as P) [kt/y] 3)	2.744 *)	1.090	2'308 /g
Nickel [t/y] 3)	29.07 *)	13.72	154'491 /g
Zinc [t/y] 3)	163.4 *)	104.6	14'927 /g
COD [kt/y] 3)	34.57 *)	18.64	99.46 /g
Lead [t/y] 3)	20.67 *)	5.166	774'411 /g
Cadmium [t/y] 3)	0.6748 *)	0.1689	23'658'407 /g
Copper [t/y] 3)	27.97 *)	21.41	61'054 /g
EPA-PAH 16 [t/y] 4)	1.029	0.2370	18'328'131 /g
Resources			
Freshwater consumption [million m^3/y] 5)	3.324 **)	3.536 **)	265.9 /m^3
Energy efficiency/scarcity: 6)			
Primary energy carriers [PJ/y]	339.9	297.6	
Consump. of renewable energy [PJ/y]	49.45 ***)	39.05	3.775 /MJ
Non-renewable prim. energy carr. [PJ/y]	290.4	258.6	4.345 /MJ
Waste			
Waste, non-hazardous [Mt/y] 7)	0.6200	0.6200	1.613 /g
Waste, hazardous [Mt/y] 7)	0.04500	0.04500	22.22 /g

*) For Croatia no appropriate data source has been available. Thus a default value based on the EU average per person has been used.

**) Due to missing data an average value per person has been deducted from Italy, Slovenia and Greece. These reference countries have been chosen to consider similar climate conditions.

***) According to the existing data-sources the calculated target value has been already exceeded.

6.2.5   Cyprus

Environmental impact	Current flow	Critical flow	Eco-factor: EF-CY-2014 (EP)
Emissions to air:			
GHG CO_2eq [Mt/y] 1)	9.300	1.220	6.248 /g
NMVOC [kt/y] 2)	14	7.700	236.1 /g
NO_x [kt/y] 2)	21	11.76	151.8 /g
SO_2 [kt/y] 2)	38	6.460	910.6 /g
PM2.5 [kt/y] 2)	2.900	1.566	1'183 /g
NH_3 [kt/y] 2)	5.800	5.220	212.9 /g
Emissions to surface water:			
Nitrogen (as N) [kt/y] 3)	2.147	1.959	559.1 /g
Phosphorus (as P) [kt/y] 3)	0.1903	0.07561	33'280 /g
Nickel [t/y] 3)	2.519	1.189	1'782'550 /g
Zinc [t/y] 3)	15.42	9.876	158'121 /g
COD [kt/y] 3)	2.361	1.273	1'457 /g
Lead [t/y] 3)	1.472	0.3678	10'875'784 /g
Cadmium [t/y] 3)	0.05338	0.01336	299'088'177 /g
Copper [t/y] 3)	2.415	1.848	707'238 /g
EPA-PAH 16 [t/y] 4)	0.1959	0.04511	96'275'535 /g
Resources			
Freshwater consumption [million m³/y] 5)	0.1990	0.06317	49.862 *) /m³
Energy efficiency/scarcity: 6)			
Primary energy carriers [PJ/y]	105.1	91.99	
Consump. of renewable energy [PJ/y]	4.446	7.845	13.00 /MJ
Non-renewable prim. energy carr. [PJ/y]	100.6	84.14	14.21 /MJ
Waste			
Waste, non-hazardous [Mt/y] 7)	0.8860	0.8860	1.129 /g
Waste, hazardous [Mt/y] 7)	0.03700	0.03700	27.03 /g

*) The relatively high value results from a water utilization index of 63%, which expresses a high water stress.

6.2.6   Czech. Republik

Environmental impact	Current flow	Critical flow	Eco-factor: EF-CZ-2014 (EP)
Emissions to air:			
GHG CO₂eq [Mt/y] 1)	131.5	39.22	0.08549 /g
NMVOC [kt/y] 2)	182	149.2	8.171 /g
NOₓ [kt/y] 2)	286	185.9	8.276 /g
SO₂ [kt/y] 2)	219	120.5	15.09 /g
PM2.5 [kt/y] 2)	22	18.26	65.98 /g
NH₃ [kt/y] 2)	82	76.26	14.10 /g
Emissions to surface water:			
Nitrogen (as N) [kt/y] 3)	139.1	126.9	8.630 /g
Phosphorus (as P) [kt/y] 3)	4.797	1.906	1'320 /g
Nickel [t/y] 3)	63.28	29.86	70'968 /g
Zinc [t/y] 3)	318.4	203.9	7'658 /g
COD [kt/y] 3)	43.91	23.68	78.32 /g
Lead [t/y] 3)	39.66	9.913	403'558 /g
Cadmium [t/y] 3)	0.9230	0.2310	17'297'146 /g
Copper [t/y] 3)	55.77	42.68	30'622 /g
EPA-PAH 16 [t/y] 4)	2.502	0.5761	7'537'993 /g
Resources			
Freshwater consumption [million m³/y] 5)	1.947	3.245	184.9 /m³
Energy efficiency/scarcity: 6)			
Primary energy carriers [PJ/y]	1'791	1'568	
Consump. of renewable energy [PJ/y]	135.9 *)	133.8	0.7357 /MJ
Non-renewable prim. energy carr. [PJ/y]	1'655	1'435	0.8043 /MJ
Waste			
Waste, non-hazardous [Mt/y] 7)	12.93	12.93	0.07736 /g
Waste, hazardous [Mt/y] 7)	1.363	1.363	0.7337 /g

*) According to the existing data-sources the calculated target value has been already exceeded.

6.2.7   Denmark

Environmental impact	Current flow	Critical flow	Eco-factor: EF-DK-2014 (EP)
Emissions to air:			
GHG CO_2eq [Mt/y] 1)	51.60	13.74	0.2733 /g
NMVOC [kt/y] 2)	110	71.50	21.52 /g
NO_x [kt/y] 2)	181	79.64	28.54 /g
SO_2 [kt/y] 2)	23	14.95	102.9 /g
PM2.5 [kt/y] 2)	25	16.75	89.11 /g
NH_3 [kt/y] 2)	83	63.08	20.86 /g
Emissions to surface water:			
Nitrogen (as N) [kt/y] 3)	181.4	165.6	6.617 /g
Phosphorus (as P) [kt/y] 3)	5.276	2.097	1'200 /g
Nickel [t/y] 3)	31.01	14.63	144'798 /g
Zinc [t/y] 3)	175.4	112.3	13'903 /g
COD [kt/y] 3)	34.52	18.62	99.61 /g
Lead [t/y] 3)	16.75	4.187	955'476 /g
Cadmium [t/y] 3)	0.5840	0.1462	27'337'523 /g
Copper [t/y] 3)	29.72	22.74	57'469 /g
EPA-PAH 16 [t/y] 4)	1.324	0.3048	14'248'758 /g
Resources			
Freshwater consumption [million m^3/y] 5)	0.6540	2.725	88.07 /m^3
Energy efficiency/scarcity: 6)			
Primary energy carriers [PJ/y]	759.5	665.0	
Consump. of renewable energy [PJ/y]	130.4	130.9	1.771 /MJ
Non-renewable prim. energy carr. [PJ/y]	629.2	534.1	2.205 /MJ
Waste			
Waste, non-hazardous [Mt/y] 7)	10.57	10.57	0.09464 /g
Waste, hazardous [Mt/y] 7)	1.338	1.338	0.7474 /g

6.2.8   Estonia

Environmental impact	Current flow	Critical flow	Eco-factor: EF-EE-2014 (EP)
Emissions to air:			
GHG CO₂eq [Mt/y] 1)	19.20	8.120	0.2912 /g
NMVOC [kt/y] 2)	41	36.90	30.11 /g
NOₓ [kt/y] 2)	36	29.52	41.31 /g
SO₂ [kt/y] 2)	76	51.68	28.46 /g
PM2.5 [kt/y] 2)	20	17	69.20 /g
NH₃ [kt/y] 2)	9.800	9.702	104.1 /g
Emissions to surface water:			
Nitrogen (as N) [kt/y] 3)	9.278	8.469	129.4 /g
Phosphorus (as P) [kt/y] 3)	0.3953	0.1571	16'020 /g
Nickel [t/y] 3)	7.364	3.475	609'780 /g
Zinc [t/y] 3)	35.77	22.91	68'170 /g
COD [kt/y] 3)	10.42	5.620	330.0 /g
Lead [t/y] 3)	14.48	3.619	1'105'458 /g
Cadmium [t/y] 3)	0.1941	0.04857	82'267'820 /g
Copper [t/y] 3)	7.246	5.545	235'702 /g
EPA-PAH 16 [t/y] 4)	0.3189	0.07342	59'149'279 /g
Resources			
Freshwater consumption [million m³/y] 5)	1.388	1.888	389.2 /m³
Energy efficiency/scarcity: 6)			
Primary energy carriers [PJ/y]	256.3	224.4	
Consump. of renewable energy [PJ/y]	44.23 *)	36.80	5.038 /MJ
Non-renewable prim. energy carr. [PJ/y]	212.0	187.6	6.027 /MJ
Waste			
Waste, non-hazardous [Mt/y] 7)	11.88 **)	11.88 **)	0.08415 /g
Waste, hazardous [Mt/y] 7)	0.2272	0.2272	4.402 /g

*) According to the existing data-sources the calculated target value has been already exceeded.

**) The Eurostat value for hazardous waste has been unreasonably high. Thus a default value based on an EU average quantity per person has been used.

6.2.9   Finland

Environmental impact	Current flow	Critical flow	Eco-factor: EF-FI-2014 (EP)
Emissions to air:			
GHG CO_2eq [Mt/y] 1)	61	14.06	0.3086 /g
NMVOC [kt/y] 2)	131	85.15	18.07 /g
NO_x [kt/y] 2)	177	115.1	13.37 /g
SO_2 [kt/y] 2)	69	48.30	29.58 /g
PM2.5 [kt/y] 2)	36	25.20	56.69 /g
NH_3 [kt/y] 2)	39	31.20	40.06 /g
Emissions to surface water:			
Nitrogen (as N) [kt/y] 3)	58.06	53.00	20.67 /g
Phosphorus (as P) [kt/y] 3)	2.193	0.8717	2'887 /g
Nickel [t/y] 3)	89.05	42.02	50'426 /g
Zinc [t/y] 3)	443.3	283.9	5'501 /g
COD [kt/y] 3)	168.0	90.58	20.47 /g
Lead [t/y] 3)	82.76	20.69	193'400 /g
Cadmium [t/y] 3)	1.369	0.3427	11'661'066 /g
Copper [t/y] 3)	63.37	48.49	26'954 /g
EPA-PAH 16 [t/y] 4)	1.280	0.2947	14'736'843 /g
Resources			
Freshwater consumption [million m^3/y] 5)	6.562	52.50	2.381 /m^3
Energy efficiency/scarcity: 6)			
Primary energy carriers [PJ/y]	1'427	1'250	
Consump. of renewable energy [PJ/y]	415.8 *)	311.5	0.8628 /MJ
Non-renewable prim. energy carr. [PJ/y]	1'011	938.1	1.149 /MJ
Waste			
Waste, non-hazardous [Mt/y] 7)	22.28	22.28	0.04488 /g
Waste, hazardous [Mt/y] 7)	2.559	2.559	0.3908 /g

*) According to the existing data-sources the calculated target value has been already exceeded.

6.2.10   France

Environmental impact	Current flow	Critical flow	Eco-factor: EF-FR-2014 (EP)
Emissions to air:			
GHG CO_2eq [Mt/y] 1)	490.1	111.5	0.03944 /g
NMVOC [kt/y] 2)	1'232	702.2	2.498 /g
NO_x [kt/y] 2)	1'430	715.0	2.797 /g
SO_2 [kt/y] 2)	467	210.2	10.57 /g
PM2.5 [kt/y] 2)	304	221.9	6.173 /g
NH_3 [kt/y] 2)	661	634.6	1.642 /g
Emissions to surface water:			
Nitrogen (as N) [kt/y] 3)	936.6	854.9	1.281 /g
Phosphorus (as P) [kt/y] 3)	30.16	11.99	210.0 /g
Nickel [t/y] 3)	377.8	178.3	11'885 /g
Zinc [t/y] 3)	2'646	1'694	921.7 /g
COD [kt/y] 3)	487.6	263.0	7.05 /g
Lead [t/y] 3)	251.3	62.82	63'685 /g
Cadmium [t/y] 3)	10.96	2.743	1'456'691 /g
Copper [t/y] 3)	366.7	280.6	4'657 /g
EPA-PAH 16 [t/y] 4)	15.46	3.561	1'219'680 /g
Resources			
Freshwater consumption [million m³/y] 5)	31.62	37.20	22.85 /m³
Energy efficiency/scarcity: 6)			
Primary energy carriers [PJ/y]	10'818	9'472	
Consump. of renewable energy [PJ/y]	869.4	1'429	0.1306 /MJ
Non-renewable prim. energy carr. [PJ/y]	9'949	8'043	0.1538 /MJ
Waste			
Waste, non-hazardous [Mt/y] 7)	82.26	82.26	0.01216 /g
Waste, hazardous [Mt/y] 7)	11.54	11.54	0.08667 /g

6.2.11   Germany (for the purpose of comparision)

Environmental impact	Current flow	Critical flow	Eco-factor*): EF-DE-2014 (EP)
Emissions to air:			
GHG CO₂eq [Mt/y]	916'769	246'486	0.015 /g
NMVOC [kt/y]	1'006	826.0	1.475 /g
NOₓ [kt/y]	1'288	652.0	3.03 /g
SO₂ [kt/y]	445.0	324.0	4.239 /g
PM2.5 [kt/y]	111.0	79.0	17.79 /g
NH₃ [kt/y]	563.0	426.0	3.102 /g
Emissions to surface water:			
Nitrogen (as N) [kt/y]	564'800	515'550	2.125 /g
Phosphorus (as P) [kt/y]	22'200	8'822	285.2 /g
Nickel [t/y]	476.8	225.0	9'418 /g
Zinc [t/y]	2'755	1'765	885 /g
COD [kt/y]	490'800	264'666	7.01 /g
Lead [t/y]	263.0	65.75	60'846 /g
Cadmium [t/y]	9.23	2.31	1'729'728 /g
Copper [t/y]	461.2	352.9	3'703 /g
EPA-PAH 16 [t/y]	19.16	4.41	985'186 /g
Resources			
Freshwater consumption [million m³/y]	32'000	37'600	22.63 /m³
Energy efficiency/scarcity:			
Primary energy carriers [PJ/y]	13'599	7'140	
Consump. of renewable energy [PJ/y]	1'463	2'245	0.349 /MJ
Non-renewable prim. energy carr. PJ/y]	12'136	4'895	0.506 /MJ
Waste			
Waste, non-hazardous [Mt/y]	136.82	136.82	0.0073 /g
Waste, hazardous [Mt/y]	15.73	15.73	0.0636 /g

*) These Eco Factors have been worked out based on the existing german environmental targets. The targets correspond with the current environmental policy of german authorities and have been coordinated with them.

Details: see Ahbe et al. 2014

6.2.12   Greece

Environmental impact	Current flow	Critical flow	Eco-factor: EF-GR-2014 (EP)
Emissions to air:			
GHG CO_2eq [Mt/y] 1)	111	20.98	0.2522 /g
NMVOC [kt/y] 2)	222	102.1	21.29 /g
NO_x [kt/y] 2)	419	289.1	5.013 /g
SO_2 [kt/y] 2)	542	140.9	27.29 /g
PM2.5 [kt/y] 2)	56	36.40	42.27 /g
NH_3 [kt/y] 2)	68	63.24	17.00 /g
Emissions to surface water:			
Nitrogen (as N) [kt/y] 3)	122.5	111.8	9.797 /g
Phosphorus (as P) [kt/y] 3)	7.913	3.145	800.2 /g
Nickel [t/y] 3)	72.24	34.09	62'166 /g
Zinc [t/y] 3)	381.5	244.3	6'391 /g
COD [kt/y] 3)	44.56	24.03	77.17 /g
Lead [t/y] 3)	35.38	8.844	452'361 /g
Cadmium [t/y] 3)	1.191	0.2981	13'404'937 /g
Copper [t/y] 3)	60.20	46.06	28'373 /g
EPA-PAH 16 [t/y] 4)	2.675	0.6159	7'051'731 /g
Resources			
Freshwater consumption [million m^3/y] 5)	9.539	14.45	45.67 /m^3
Energy efficiency/scarcity: 6)			
Primary energy carriers [PJ/y]	1'162	1'017	
Consump. of renewable energy [PJ/y]	95.23	120.1	1.169 /MJ
Non-renewable prim. energy carr. [PJ/y]	1'067	897.1	1.325 /MJ
Waste			
Waste, non-hazardous [Mt/y] 7)	23.41	23.41	0.04271 /g
Waste, hazardous [Mt/y] 7)	0.2530	0.2530	3.953 /g

6.2.13 Hungary

Environmental impact	Current flow	Critical flow	Eco-factor: EF-HU-2014 (EP)
Emissions to air:			
GHG CO$_2$eq [Mt/y] 1)	62	19.52	0.1627 /g
NMVOC [kt/y] 2)	177	123.9	11.53 /g
NO$_x$ [kt/y] 2)	203	134.0	11.31 /g
SO$_2$ [kt/y] 2)	129	69.66	26.58 /g
PM2.5 [kt/y] 2)	31	26.97	42.62 /g
NH$_3$ [kt/y] 2)	80	72	15.43 /g
Emissions to surface water:			
Nitrogen (as N) [kt/y] 3)	84.52	77.15	14.20 /g
Phosphorus (as P) [kt/y] 3)	4.616	1.834	1'372 /g
Nickel [t/y] 3)	81.23	38.33	55'283 /g
Zinc [t/y] 3)	326.0	208.8	7'479 /g
COD [kt/y] 3)	49.86	26.89	68.97 /g
Lead [t/y] 3)	50.95	12.73	314'146.83 /g
Cadmium [t/y] 3)	1.207	0.3021	13'227'165 /g
Copper [t/y] 3)	55.07	42.14	31'015 /g
EPA-PAH 16 [t/y] 4)	2.395	0.5515	7'875'034 /g
Resources			
Freshwater consumption [million m^3/y] 5)	6.070	20.93	13.86 /m^3
Energy efficiency/scarcity: 6)			
Primary energy carriers [PJ/y]	986.0	863.3	
Consump. of renewable energy [PJ/y]	82.17 *)	73.63	1.326 /MJ
Non-renewable prim. energy carr. [PJ/y]	903.9	789.7	1.449 /MJ
Waste			
Waste, non-hazardous [Mt/y] 7)	12.04	12.04	0.08309 /g
Waste, hazardous [Mt/y] 7)	0.5410	0.5410	1.848 /g

*) According to the existing data-sources the calculated target value has been already exceeded.

6.2.14 Ireland

Environmental impact	Current flow	Critical flow	Eco-factor: EF-IE-2014 (EP)
Emissions to air:			
GHG CO_2eq [Mt/y] 1)	58.50	11.04	0.4800 /g
NMVOC [kt/y] 2)	57	42.75	31.19 /g
NO_x [kt/y] 2)	127	64.77	30.27 /g
SO_2 [kt/y] 2)	71	24.85	115.0 /g
PM2.5 [kt/y] 2)	11	9.020	135.2 /g
NH_3 [kt/y] 2)	109	107.9	9.361 /g
Emissions to surface water:			
Nitrogen (as N) [kt/y] 3)	385.3	351.7	3.115 /g
Phosphorus (as P) [kt/y] 3)	10.50	4.175	602.8 /g
Nickel [t/y] 3)	25.87	12.21	173'597 /g
Zinc [t/y] 3)	169.6	108.6	14'382 /g
COD [kt/y] 3)	30.02	16.19	114.5 /g
Lead [t/y] 3)	41.16	10.29	388'829 /g
Cadmium [t/y] 3)	0.4825	0.1207	33'090'736 /g
Copper [t/y] 3)	23.83	18.23	71'677 /g
EPA-PAH 16 [t/y] 4)	1.088	0.2505	17'334'738 /g
Resources			
Freshwater consumption [million m^3/y] 5)	16.88	168.8	0.5923 /m^3
Energy efficiency/scarcity: 6)			
Primary energy carriers [PJ/y]	579.8	507.6	
Consump. of renewable energy [PJ/y]	31.15	53.28	2.379 /MJ
Non-renewable prim. energy carr. [PJ/y]	548.6	454.3	2.658 /MJ
Waste			
Waste, non-hazardous [Mt/y] 7)	14.03	14.03	0.07128 /g
Waste, hazardous [Mt/y] 7)	1.972	1.972	0.5071 /g

6.2.15   Italy

Environmental impact	Current flow	Critical flow	Eco-factor: EF-IT-2014 (EP)
Emissions to air:			
GHG CO_2eq [Mt/y] 1)	460.1	103.8	0.04269 /g
NMVOC [kt/y] 2)	1'286	835.9	1.840 /g
NO_x [kt/y] 2)	1'212	727.2	2.292 /g
SO_2 [kt/y] 2)	403	262.0	5.873 /g
PM2.5 [kt/y] 2)	166	149.4	7.437 /g
NH_3 [kt/y] 2)	416	395.2	2.664 /g
Emissions to surface water:			
Nitrogen (as N) [kt/y] 3)	679.1	619.9	1.767 /g
Phosphorus (as P) [kt/y] 3)	55.01	21.86	115.1 /g
Nickel [t/y] 3)	563.4	265.9	7'971 /g
Zinc [t/y] 3)	2'570	1'646	948.8 /g
COD [kt/y] 3)	454.4	245.0	7.57 /g
Lead [t/y] 3)	321.2	80.30	49'821 /g
Cadmium [t/y] 3)	11.78	2.948	1'355'175 /g
Copper [t/y] 3)	374.8	286.8	4'557 /g
EPA-PAH 16 [t/y] 4)	14.16	3.260	1'332'370 /g
Resources			
Freshwater consumption [million m^3/y] 5)	44.41	37.01	32.43 /m^3
Energy efficiency/scarcity: 6)			
Primary energy carriers [PJ/y]	6'834	5'983	
Consump. of renewable energy [PJ/y]	756.0 *)	667.3	0.1911 /MJ
Non-renewable prim. energy carr. [PJ/y]	6'078	5'316	0.2151 /MJ
Waste			
Waste, non-hazardous [Mt/y] 7)	101.4	101.4	0.00986 /g
Waste, hazardous [Mt/y] 7)	6.655	6.655	0.1503 /g

*) According to the existing data-sources the calculated target value has been already exceeded.

6.2.16   Latvia

Environmental impact	Current flow	Critical flow	Eco-factor: EF-LV-2014 (EP)
Emissions to air:			
GHG CO_2eq [Mt/y] 1)	11	5.240	0.4006 /g
NMVOC [kt/y] 2)	73	53.29	25.71 /g
NO_x [kt/y] 2)	37	25.16	58.45 /g
SO_2 [kt/y] 2)	6.700	6.164	176.3 /g
PM2.5 [kt/y] 2)	27	22.68	52.49 /g
NH_3 [kt/y] 2)	16	15.84	63.77 /g
Emissions to surface water:			
Nitrogen (as N) [kt/y] 3)	60.57	55.29	19.81 /g
Phosphorus (as P) [kt/y] 3)	0.7615	0.3026	8'316 /g
Nickel [t/y] 3)	7.149	3.374	628'107 /g
Zinc [t/y] 3)	48.77	31.23	50'005 /g
COD [kt/y] 3)	12.06	6.504	285.1 /g
Lead [t/y] 3)	10.79	2.697	1'483'362 /g
Cadmium [t/y] 3)	0.1903	0.04762	83'909'553 /g
Copper [t/y] 3)	7.934	6.071	215'283 /g
EPA-PAH 16 [t/y] 4)	0.5072	0.1168	37'190'754 /g
Resources			
Freshwater consumption [million m^3/y] 5)	0.2110	4.220	11.85 /m^3
Energy efficiency/scarcity: 6)			
Primary energy carriers [PJ/y]	190.0	166.3	
Consump. of renewable energy [PJ/y]	97.61 *)	43.65	4.526 /MJ
Non-renewable prim. energy carr. [PJ/y]	92.37	122.7	6.137 /MJ
Waste			
Waste, non-hazardous [Mt/y] 7)	1.413	1.413	0.7077 /g
Waste, hazardous [Mt/y] 7)	0.06700	0.06700	14.93 /g

*) According to the existing data-sources the calculated target value has been already exceeded.

6.2.17   Lithuania

Environmental impact	Current flow	Critical flow	Eco-factor: EF-LT-2014 (EP)
Emissions to air:			
GHG CO$_2$eq [Mt/y] 1)	21.60	9.740	0.2277 /g
NMVOC [kt/y] 2)	84	57.12	25.75 /g
NO$_x$ [kt/y] 2)	58	30.16	63.76 /g
SO$_2$ [kt/y] 2)	44	19.80	112.2 /g
PM2.5 [kt/y] 2)	8.700	6.960	179.6 /g
NH$_3$ [kt/y] 2)	39	35.10	31.66 /g
Emissions to surface water:			
Nitrogen (as N) [kt/y] 3)	58.87	53.74	20.39 /g
Phosphorus (as P) [kt/y] 3)	1.689	0.6713	3'749 /g
Nickel [t/y] 3)	11.30	5.332	397'419 /g
Zinc [t/y] 3)	71.19	45.59	34'253 /g
COD [kt/y] 3)	8.872	4.784	387.6 /g
Lead [t/y] 3)	10.53	2.632	1'519'936 /g
Cadmium [t/y] 3)	0.2790	0.06983	57'221'352 /g
Copper [t/y] 3)	9.752	7.462	175'140 /g
EPA-PAH 16 [t/y] 4)	0.7515	0.1730	25'099'855 /g
Resources			
Freshwater consumption [million m^3/y] 5)	2.381	5.013	94.76 /m^3
Energy efficiency/scarcity: 6)			
Primary energy carriers [PJ/y]	296.6	259.7	
Consump. of renewable energy [PJ/y]	50.15 *)	39.19	4.304 /MJ
Non-renewable prim. energy carr. [PJ/y]	246.5	220.5	5.068 /MJ
Waste			
Waste, non-hazardous [Mt/y] 7)	5.109	5.109	0.1957 /g
Waste, hazardous [Mt/y] 7)	0.1100	0.1100	9.091 /g

*) According to the existing data-sources the calculated target value has been already exceeded.

6.2.18   Luxembourg

Environmental impact	Current flow	Critical flow	Eco-factor: EF-LU-2014 (EP)
Emissions to air:			
GHG CO_2eq [Mt/y] 1)	11.80	2.580	1.773 /g
NMVOC [kt/y] 2)	9.800	6.958	202.4 /g
NO_x [kt/y] 2)	19	10.83	162.0 /g
SO_2 [kt/y] 2)	2.500	1.650	918.3 /g
PM2.5 [kt/y] 2)	3.100	2.635	446.5 /g
NH_3 [kt/y] 2)	5	4.950	204.1 /g
Emissions to surface water:			
Nitrogen (as N) [kt/y] 3)	5.921	5.405	202.7 /g
Phosphorus (as P) [kt/y] 3)	0.2185	0.08683	28'980 /g
Nickel [t/y] 3)	3.587	1.692	1'252'074 /g
Zinc [t/y] 3)	19.58	12.54	124'540 /g
COD [kt/y] 3)	3.914	2.110	878.7 /g
Lead [t/y] 3)	2.029	0.5071	7'888'781 /g
Cadmium [t/y] 3)	0.05540	0.01387	288'182'315 /g
Copper [t/y] 3)	3.810	2.915	448'329 /g
EPA-PAH 16 [t/y] 4)	0.1201	0.02765	157'077'240 /g
Resources			
Freshwater consumption [million m^3/y] 5)	0.04700	0.2350	851.1 /m^3
Energy efficiency/scarcity: 6)			
Primary energy carriers [PJ/y]	186.5	163.3	
Consump. of renewable energy [PJ/y]	3.923	11.78	7.380 /MJ
Non-renewable prim. energy carr. [PJ/y]	182.6	151.5	7.954 /MJ
Waste			
Waste, non-hazardous [Mt/y] 7)	1.312	1.312	0.7622 /g
Waste, hazardous [Mt/y] 7)	0.3790	0.3790	2.639 /g

6.2.19   Malta

Environmental impact	Current flow	Critical flow	Eco-factor: EF-MT-2014 (EP)
Emissions to air:			
GHG CO_2eq [Mt/y] 1)	3.100	0.4000	19.38 /g
NMVOC [kt/y] 2)	3.300	2.541	511.1 /g
NO_x [kt/y] 2)	9.300	5.394	319.6 /g
SO_2 [kt/y] 2)	11	2.530	1'719 /g
PM2.5 [kt/y] 2)	1.300	0.9750	1'368 /g
NH_3 [kt/y] 2)	1.600	1.536	678.2 /g
Emissions to surface water: *)			
Nitrogen (as N) [kt/y] 3)	5.356	4.889	224.1 /g
Phosphorus (as P) [kt/y] 3)	0.2749	0.1092	23'038 /g
Nickel [t/y] 3)	2.912	1.374	1'542'326 /g
Zinc [t/y] 3)	16.36	10.48	149'017 /g
COD [kt/y] 3)	3.463	1.868	993.0 /g
Lead [t/y] 3)	2.070	0.5175	7'731'145 /g
Cadmium [t/y] 3)	0.06760	0.01692	236'187'865 /g
Copper [t/y] 3)	2.802	2.144	609'520 /g
EPA-PAH 16 [t/y] 4)	0.09902	0.02280	190'478'258 /g
Resources			
Freshwater consumption [million m^3/y] 5)	0.03100	0.02818	39'032 /m^3
Energy efficiency/scarcity: 6)			
Primary energy carriers [PJ/y]	35.06	30.69	
Consump. of renewable energy [PJ/y]	0.2596	2.014	39.53 /MJ
Non-renewable prim. energy carr. [PJ/y]	34.80	28.68	42.30 /MJ
Waste			
Waste, non-hazardous [Mt/y] 7)	0.2820	0.2820	3.546 /g
Waste, hazardous [Mt/y] 7)	0.01700	0.01700	58.82 /g

*) Due to missing data for Malta default values have been calculated based on an EU average per person.

6.2.20   *Netherlands*

Environmental impact	Current flow	Critical flow	Eco-factor: EF-NL-2014 (EP)
Emissions to air:			
GHG CO_2eq [Mt/y] 1)	191.7	42.36	0.1068 /g
NMVOC [kt/y] 2)	182	167.4	6.492 /g
NO_x [kt/y] 2)	370	203.5	8.935 /g
SO_2 [kt/y] 2)	65	46.80	29.68 /g
PM2.5 [kt/y] 2)	21	13.23	120.0 /g
NH_3 [kt/y] 2)	141	122.7	9.370 /g
Emissions to surface water:			
Nitrogen (as N) [kt/y] 3)	240.4	219.4	4.993 /g
Phosphorus (as P) [kt/y] 3)	6.142	2.441	1'031 /g
Nickel [t/y] 3)	74.07	34.95	60'624 /g
Zinc [t/y] 3)	512.9	328.4	4'755 /g
COD [kt/y] 3)	103.9	56.01	33.11 /g
Lead [t/y] 3)	63.57	15.89	251'755 /g
Cadmium [t/y] 3)	1.748	0.4374	9'135'774 /g
Copper [t/y] 3)	68.92	52.74	24'781 /g
EPA-PAH 16 [t/y] 4)	3.964	0.9128	4'757'960 /g
Resources			
Freshwater consumption [million m³/y] 5)	10.83	19.68	27.94 /m³
Energy efficiency/scarcity: 6)			
Primary energy carriers [PJ/y]	3'424	2'998	
Consump. of renewable energy [PJ/y]	158.2	275.3	0.4001 /MJ
Non-renewable prim. energy carr. [PJ/y]	3'266	2'722	0.4406 /MJ
Waste			
Waste, non-hazardous [Mt/y] 7)	36.06	36.06	0.02773 /g
Waste, hazardous [Mt/y] 7)	4.565	4.565	0.2191 /g

6.2.21   Poland

Environmental impact	Current flow	Critical flow	Eco-factor: EF-PL-2014 (EP)
Emissions to air:			
GHG CO_2eq [Mt/y] 1)	399.3	93.28	0.04589 /g
NMVOC [kt/y] 2)	593	444.8	2.998 /g
NO_x [kt/y] 2)	866	606.2	2.357 /g
SO_2 [kt/y] 2)	1'224	501.8	4.860 /g
PM2.5 [kt/y] 2)	133	111.7	10.66 /g
NH_3 [kt/y] 2)	270	267.3	3.779 /g
Emissions to surface water:			
Nitrogen (as N) [kt/y] 3)	469.2	428.3	2.558 /g
Phosphorus (as P) [kt/y] 3)	27.84	11.06	227.5 /g
Nickel [t/y] 3)	431.6	203.7	10'404 /g
Zinc [t/y] 3)	2'559	1'639	952.8 /g
COD [kt/y] 3)	422.1	227.6	8.15 /g
Lead [t/y] 3)	394.4	98.60	40'576 /g
Cadmium [t/y] 3)	9.922	2.483	1'609'052 /g
Copper [t/y] 3)	308.9	236.4	5'529 /g
EPA-PAH 16 [t/y] 4)	9.128	2.102	2'066'247 /g
Resources			
Freshwater consumption [million m^3/y] 5)	11.52	12.80	70.33 /m^3
Energy efficiency/scarcity: 6)			
Primary energy carriers [PJ/y]	4'102	3'592	
Consump. of renewable energy [PJ/y]	355.0 *)	353.4	0.3222 /MJ
Non-renewable prim. energy carr. [PJ/y]	3'747	3'238	0.3574 /MJ
Waste			
Waste, non-hazardous [Mt/y] 7)	75.60	75.60	0.01323 /g
Waste, hazardous [Mt/y] 7)	1.492	1.492	0.6702 /g

*) According to the existing data-sources the calculated target value has been already exceeded.

6.2.22   Portugal

Environmental impact	Current flow	Critical flow	Eco-factor: EF-PT-2014 (EP)
Emissions to air:			
GHG CO_2eq [Mt/y] 1)	68.80	12.16	0.4653 /g
NMVOC [kt/y] 2)	207	169.7	7.185 /g
NO_x [kt/y] 2)	256	163.8	9.537 /g
SO_2 [kt/y] 2)	177	65.49	41.27 /g
PM2.5 [kt/y] 2)	65	55.25	21.29 /g
NH_3 [kt/y] 2)	50	46.50	23.12 /g
Emissions to surface water:			
Nitrogen (as N) [kt/y] 3)	49.11	44.83	24.44 /g
Phosphorus (as P) [kt/y] 3)	4.819	1.915	1'314 /g
Nickel [t/y] 3)	140.0	66.07	32'072 /g
Zinc [t/y] 3)	328.1	210.1	7'433 /g
COD [kt/y] 3)	139.3	75.10	24.69 /g
Lead [t/y] 3)	131.3	32.83	121'860 /g
Cadmium [t/y] 3)	3.322	0.8314	4'805'945 /g
Copper [t/y] 3)	77.36	59.19	22'078 /g
EPA-PAH 16 [t/y] 4)	2.529	0.5823	7'458'580 /g
Resources			
Freshwater consumption [million m^3/y] 5)	8.289	11.05	67.86 /m^3
Energy efficiency/scarcity: 6)			
Primary energy carriers [PJ/y]	929.5	813.8	
Consump. of renewable energy PJ/y]	182.4 *)	165.5	1.416 /MJ
Non-renewable prim. energy carr. [PJ/y]	747.0	648.3	1.777 /MJ
Waste			
Waste, non-hazardous [Mt/y] 7)	24.45	24.45	0.0409 /g
Waste, hazardous [Mt/y] 7)	1.625	1.625	0.6154 /g

*) According to the existing data-sources the calculated target value has been already exceeded.

6.2.23   Romania

Environmental impact	Current flow	Critical flow	Eco-factor: EF-RO-2014 (EP)
Emissions to air:			
GHG CO$_2$eq [Mt/y] 1)	118.8	49.54	0.0484 /g
NMVOC [kt/y] 2)	425	318.8	4.183 /g
NO$_x$ [kt/y] 2)	309	170.0	10.70 /g
SO$_2$ [kt/y] 2)	643	147.9	29.40 /g
PM2.5 [kt/y] 2)	106	76.32	18.20 /g
NH$_3$ [kt/y] 2)	199	173.1	6.639 /g
Emissions to surface water:			
Nitrogen (as N) [kt/y] 3)	235.6	215.1	5.094 /g
Phosphorus (as P) [kt/y] 3)	18.23	7.243	347.4 /g
Nickel [t/y] 3)	76.15	35.93	58'973 /g
Zinc [t/y] 3)	394.3	252.5	6'184 /g
COD [kt/y] 3)	114.2	61.57	30.12 /g
Lead [t/y] 3)	80.21	20.05	199'542 /g
Cadmium [t/y] 3)	2.606	0.6521	6'127'371 /g
Copper [t/y] 3)	99.50	76.13	17'166 /g
EPA-PAH 16 [t/y] 4)	4.854	1.118	3'885'902 /g
Resources			
Freshwater consumption [million m^3/y] 5)	6.219	44.42	3.152 /m^3
Energy efficiency/scarcity: 6)			
Primary energy carriers [PJ/y]	1'481	1'297	
Consump. of renewable energy [PJ/y]	219.4 *)	204.2	0.8905 /MJ
Non-renewable prim. energy carr. [PJ/y]	1'261	1'092	1.057 /MJ
Waste			
Waste, non-hazardous [Mt/y] 7)	40.45	40.45	0.02472 /g
Waste, hazardous [Mt/y] 7)	0.7030	0.7030	1.422 /g

*) According to the existing data-sources the calculated target value has been already exceeded.

6.2.24   Slovakia

Environmental impact	Current flow	Critical flow	Eco-factor: EF-SK-2014 (EP)
Emissions to air:			
GHG CO_2eq [Mt/y] 1)	42.70	14.64	0.1992 /g
NMVOC [kt/y] 2)	73	59.86	20.37 /g
NO_x [kt/y] 2)	102	65.28	23.94 /g
SO_2 [kt/y] 2)	89	38.27	60.77 /g
PM2.5 [kt/y] 2)	37	23.68	65.98 /g
NH_3 [kt/y] 2)	29	24.65	47.73 /g
Emissions to surface water:			
Nitrogen (as N) [kt/y] 3)	43.05	39.29	27.88 /g
Phosphorus (as P) [kt/y] 3)	2.481	0.9859	2'553 /g
Nickel [t/y] 3)	22.32	10.53	201'164 /g
Zinc [t/y] 3)	111.0	71.07	21'971 /g
COD [kt/y] 3)	36.53	19.70	94.15 /g
Lead [t/y] 3)	15.58	3.895	1'026'997 /g
Cadmium [t/y] 3)	0.5190	0.1299	30'761'587 /g
Copper [t/y] 3)	20.14	15.41	84'800 /g
EPA-PAH 16 [t/y] 4)	1.289	0.2968	14'630'268 /g
Resources			
Freshwater consumption [million m^3/y] 5)	1.258	16.77	4.471 /m^3
Energy efficiency/scarcity: 6)			
Primary energy carriers [PJ/y]	699.3	612.2	
Consump. of renewable energy [PJ/y]	60.02 *)	56.23	1.878 /MJ
Non-renewable prim. energy carr. [PJ/y]	639.2	556.0	2.068 /MJ
Waste			
Waste, non-hazardous [Mt/y] 7)	8.156	8.156	0.1226 /g
Waste, hazardous [Mt/y] 7)	0.4370	0.4370	2.288 /g

*) According to the existing data-sources the calculated target value has been already exceeded.

6.2.25   Slovenia

Environmental impact	Current flow	Critical flow	Eco-factor: EF-SI-2014 (EP)
Emissions to air:			
GHG CO$_2$eq [Mt/y] 1)	18.90	3.680	1.396 /g
NMVOC [kt/y] 2)	37	28.49	45.58 /g
NO$_x$ [kt/y] 2)	47	28.67	57.18 /g
SO$_2$ [kt/y] 2)	40	14.80	182.6 /g
PM2.5 [kt/y] 2)	14	10.50	127.0 /g
NH$_3$ [kt/y] 2)	18	17.82	56.68 /g
Emissions to surface water:			
Nitrogen (as N) [kt/y] 3)	38.75	35.37	30.98 /g
Phosphorus (as P) [kt/y] 3)	4.406	1.751	1'437 /g
Nickel [t/y] 3)	11.06	5.219	406'036 /g
Zinc [t/y] 3)	55.86	35.77	43'654 /g
COD [kt/y] 3)	9.605	5.179	358.0 /g
Lead [t/y] 3)	6.700	1.675	2'388'653 /g
Cadmium [t/y] 3)	0.1969	0.04928	81'074'673 /g
Copper [t/y] 3)	9.371	7.170	182'265 /g
EPA-PAH 16 [t/y] 4)	0.4896	0.1127	38'526'657 /g
Resources			
Freshwater consumption [million m^3/y] 5)	1.058	7.053	21.27 /m^3
Energy efficiency/scarcity: 6)			
Primary energy carriers [PJ/y]	293.3	256.8	
Consump. of renewable energy [PJ/y]	41.43	42.11	4.569 /MJ
Non-renewable prim. energy carr. [PJ/y]	251.8	214.7	5.466 /MJ
Waste			
Waste, non-hazardous [Mt/y] 7)	3.458	3.458	0.2892 /g
Waste, hazardous [Mt/y] 7)	0.1170	0.1170	8.547 /g

6.2.26   Spain

Environmental impact	Current flow	Critical flow	Eco-factor: EF-ES-2014 (EP)	
Emissions to air:				
GHG CO_2eq [Mt/y] 1)	340.8	56.74	0.1059	/g
NMVOC [kt/y] 2)	809	631.0	2.032	/g
NO_x [kt/y] 2)	1'292	762.3	2.223	/g
SO_2 [kt/y] 2)	1'282	423.1	7.163	/g
PM2.5 [kt/y] 2)	93	79.05	14.88	/g
NH_3 [kt/y] 2)	365	354.1	2.912	/g
Emissions to surface water:				
Nitrogen (as N) [kt/y] 3)	352.7	321.9	3.403	/g
Phosphorus (as P) [kt/y] 3)	27.49	10.92	230.4	/g
Nickel [t/y] 3)	344.1	162.4	13'050	/g
Zinc [t/y] 3)	1'646	1'054	1'481	/g
COD [kt/y] 3)	403.1	217.3	8.53	/g
Lead [t/y] 3)	190.2	47.55	84'131	/g
Cadmium [t/y] 3)	8.124	2.033	1'965'205	/g
Copper [t/y] 3)	307.8	235.5	5'549	/g
EPA-PAH 16 [t/y] 4)	11.10	2.560	1'693'726	/g
Resources				
Freshwater consumption [million m^3/y] 5)	32.47	20.95	74.00	/m^3
Energy efficiency/scarcity: 6)				
Primary energy carriers [PJ/y]	5'330	4'666		
Consump. of renewable energy [PJ/y]	606.6	612.3	0.2497	/MJ
Non-renewable prim. energy carr. [PJ/y]	4'723	4'054	0.2874	/MJ
Waste				
Waste, non-hazardous [Mt/y] 7)	64.85	64.85	0.01542	/g
Waste, hazardous [Mt/y] 7)	2.991	2.991	0.3343	/g

6.2.27   Sweden

Environmental impact	Current flow	Critical flow	Eco-factor: EF-SW-2014 (EP)
Emissions to air:			
GHG CO_2eq [Mt/y] 1)	57.60	14.54	0.2725 /g
NMVOC [kt/y] 2)	197	147.8	9.024 /g
NO_x [kt/y] 2)	174	111.4	14.03 /g
SO_2 [kt/y] 2)	36	28.08	45.66 /g
PM2.5 [kt/y] 2)	29	23.49	52.56 /g
NH_3 [kt/y] 2)	55	46.75	25.17 /g
Emissions to surface water:			
Nitrogen (as N) [kt/y] 3)	99.84	91.13	12.02 /g
Phosphorus (as P) [kt/y] 3)	2.218	0.8814	2'855 /g
Nickel [t/y] 3)	25.00	11.80	179'645 /g
Zinc [t/y] 3)	382.0	244.6	6'383 /g
COD [kt/y] 3)	209.8	113.1	16.39 /g
Lead [t/y] 3)	94.89	23.72	168'672 /g
Cadmium [t/y] 3)	1.212	0.3034	13'171'262 /g
Copper [t/y] 3)	24.19	18.51	70'600 /g
EPA-PAH 16 [t/y] 4)	2.234	0.5144	8'442'975 /g
Resources			
Freshwater consumption [million m^3/y] 5)	2.630	29.22	3.080 /m^3
Energy efficiency/scarcity: 6)			
Primary energy carriers [PJ/y]	2'085	1'825	
Consump. of renewable energy [PJ/y]	775.0 *)	586.8	0.5794 /MJ
Non-renewable prim. energy carr. [PJ/y]	1'310	1'239	0.8539 /MJ
Waste			
Waste, non-hazardous [Mt/y] 7)	16.70	16.70	0.0599 /g
Waste, hazardous [Mt/y] 7)	2.515	2.515	0.3976 /g

*) According to the existing data-sources the calculated target value has been already exceeded.

6.2.28   United Kingdom

Environmental impact	Current flow	Critical flow	Eco-factor: EF-GB-2014 (EP)
Emissions to air:			
GHG CO₂eq [Mt/y] 1)	580.8	155.1	0.02414 /g
NMVOC [kt/y] 2)	1'088	739.8	1.988 /g
NOₓ [kt/y] 2)	1'580	711.0	3.125 /g
SO₂ [kt/y] 2)	706	289.5	8.426 /g
PM2.5 [kt/y] 2)	81	56.70	25.20 /g
NH₃ [kt/y] 2)	307	282.4	3.848 /g
Emissions to surface water:			
Nitrogen (as N) [kt/y] 3)	1'202	1'097	0.9985 /g
Phosphorus (as P) [kt/y] 3)	66.60	26.46	95.09 /g
Nickel [t/y] 3)	352.6	166.4	12'736 /g
Zinc [t/y] 3)	1'817	1'163	1'342 /g
COD [kt/y] 3)	627.0	338.1	5.48 /g
Lead [t/y] 3)	224.6	56.14	71'258 /g
Cadmium [t/y] 3)	8.401	2.103	1'900'379 /g
Copper [t/y] 3)	653.6	500.1	2'613 /g
EPA-PAH 16 [t/y] 4)	14.95	3.442	1'261'604 /g
Resources			
Freshwater consumption [million m³/y] 5)	23.18 *)	37.70 *)	16.31 /m³
Energy efficiency/scarcity: 6)			
Primary energy carriers [PJ/y]	8'470	7'416	
Consump. of renewable energy [PJ/y]	297.1	1'112	0.1748 /MJ
Non-renewable prim. energy carr. [PJ/y]	8'172	6'303	0.2057 /MJ
Waste			
Waste, non-hazardous [Mt/y] 7)	139.9	139.9	0.00715 /g
Waste, hazardous [Mt/y] 7)	7.285	7.285	0.1373 /g

*) The data source delivers values only for England and Wales. They have been extrapolated according to the population ratio.

7 Environmental Impact Calculation

To give a short introduction into an environmental impact assessment a hypothetical example will be calcuated as a comparison between two processes. Both of them are designed to fulfill the same job but they take different workstep approaches. Within the environmental impact assessment the focus is on the environmental outcome for every process. The inventory analysis for both processes is shown in table 39.

Table 39: Inventory analysis of both processes (quantity per year)

Impact	Unit	Process 1	Process 2
CO_2-eq	Kg	582.000	300.000
NMVOC	Kg	4.100	0
Waste, non hazardous	Kg	138.000	0
Consumption of non renewable energy*	MJ	5.460.000	0
Consumption of renewable energy*	MJ	0	4.200.000
COD	Kg	0	2.600
Consumption of freshwater	m3	0	62.000

* to keep the example simple one energy source instead of an energymix is assumed

After the inventory analysis is clearly defined each corresponding ecofactor has to be assigned correctly (german ecofactors are used in this example). Throughout a simple multiplication between ecofactor and quantity parameter of the inventory analysis the ecopoints (EP) can be calculated. By repeating this procedure for every impact and adding up of the EPs the final overall environmental impacts of the two processes are optained. The calculation and the results are shown in table 43.

Table 43: Calculation of the ecopoints

Impact	Unit	EF [EP/unit]	Invent. proc. 1	Invent. proc. 2	EP process 1	EP prozess 2
CO2-eq	Kg	15	582.000	300.000	8.730.000	4.500.000
NMVOC	Kg	1475	4.100	0	6.047.500	0
Abfall, non hazardous	Kg	7,3	138.000	0	1.007.400	0
Consumption of non renewable Energie	MJ	0,506	5.460.000	0	2.762.760	0
Consumption of renewable energy	MJ	0,349	0	4.200.000	0	1.465.800
COD	Kg	7010	0	2.600	0	18.226.000
consumption drinking water	m³	22,63	0	62.000	0	1.403.060
				Sum	18.547.660	25.594.860

The result of both analysed processes is clearly represented in figure 5 as sums of the ecopoints. In addition the influences of the single impacts are to be seen.

In this comparison it is clearly to be seen that process no. 2 shows a higher environmental load with 25.6 mio EP/y than process no. 1 with 18,5 Mio EP/y. The graphic representation provides a deeper insight into the influences of the single impacts for the overall result. Thus this divergent view allows a very specific analytical review of the calculation and the given results.

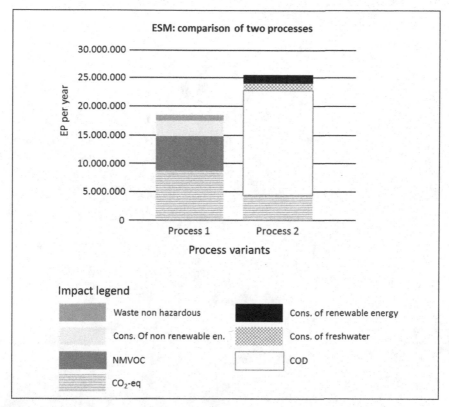

Figure 5: Results

8 Sources

Ahbe, S., Braunschweig, A., Müller-Wenk, R. (1990): „Methodik für Ökobilanzen", environment publication series, no. 133 (publ. SAEFL, Berne).

Ahbe, S., Schebek, L., Jansky, N., Wellge, S., Weihofen, S. (2014): Methode der ökologischen Knappheit für Deutschland – eine Initiative der Volkswagen AG, Logos-Verlag Berlin GmbH, Berlin.

BMWi and BMU. (2012). Erster Monitoring-Bericht „Energie der Zukunft". Berlin: Federal Ministry of Economics and Technology (BMWi) and Federal Ministry for the Environment, Nature Conservation and Nuclear Safety (BMU).

Brown & Matlock, 2011, "A Review of Water Scarcity Indices and methodologies", University of Arkansas, The Sustainability Consortium

Deltares (2013). Van den Roovaart et al.: Diffuse Water Emissions in E-PRTR, July 2013, Project report for the European Commission, http://prtr.ec.europa.eu/docs/water_report.pdf, retrieved on 2.3.2015

German Federal Government. (2010). Energiekonzept für eine umweltschonende, zuverlässige und bezahlbare Energieversorgung. Berlin: German Federal Government.

German Federal Government. (2013). Glossar zur Energie. Berlin: German Federal Government: http://www.bundesregierung.de/Content/DE/StatischeSeiten/Breg/FAQ/faq-energie.html

Deutsches Institut für Normung e.V. (1980). DIN 38 409, part 41: Summary Action and Material Characteristic Parameters (Group H); Determination of the Chemical Oxygen Demand (COD) Berlin: Beuth Verlag GmbH.

Deutsches Institut für Normung e.V. (2006). DIN EN ISO 14040: Environmental management – Life cycle assessment – Principles and framework. Berlin: Beuth Verlag GmbH.

ECE/EB.AIR/114, 2012, Gothenburg Protocol, 2005 emission level and national emission reduction commitments in 2020 and beyond http://www.unece.org/fileadmin/DAM/press/pr2012/GothenburgProtocol_Table_Eng.pdf, retrieved on 2.3.2015

EEA (2009) self-description based on (EC) no. 401/2009 on the European Environment Agency and the European Environment Information and Observation Network: http://europa.eu/ybout-eu/agencies/regulatory_agencies_bodies/policy_agencies /eea/index_de.htm; retrieved on 26.02.2015

EEA (2010) European Environment Agency: Water Exploitation Index (WEI), published 1.12.2010; http://www.eea.europa.eu/data-and-maps/figures/water-exploitation-index-wei-3, retrieved on 3.12.2014.

EEA (2012) European Environment Agency: Freshwater Abstraction and Hydropower, published 9.7.2012; http://www.eea.europa.eu/data-and-maps/data/freshwater-abstraction-and-hydropower-2013, retrieved on 3.12.2014

EEA (2013) European Environment Agency: Towards a green economy in Europe; EU environmental policy targets and objectives 2010–2050, EEA Report No. 8/2013

EEA (2014a) European Environment Agency: Personal communication by email with Mr. Martin Adams (EEA) about "data source for EU emissions to air" of 10.7.2014

EEA (2014b) European Environment Agency: Personal communication by email with Mr. Bo Jacobsen (EEA) about "EU targets for emissions to surface water" of 16.3.2014.

EEA (2014c) European Environment Agency: Personal communication by email with Mrs. Almut Reichel (EEA) about "EU targets for waste generation and treatment" of 3.7.2014.

EEA (2014d) European Environment Agency: Personal communication by email with Mr. Bo Jacobsen (EEA) about "EU targets for emissions to surface water" of 02.10.2014

EU (2006) Regulation (EC) No 166/2006 of the European Parliament and of the Council of 18 January 2006 concerning the establishment of a European Pollutant Release and Transfer Register and amending Council Directives 91/689/EEC and 96/61/EC

European Commission. (2005). Communication from the Commission to the Council and the European Parliament of 21 September 2005 "Thematic Strategy on air pollution" COM(2005) 446.

European Environment Agency (2014a): E-PRTR The European Pollutant Release and Transfer Register, http://prtr.ec.europa.eu/pgAbout.aspx, retrieved on 2.3.2015

European Environment Agency EEA (2014): Annual European Community greenhouse gas inventory 1990–2012 and inventory report 2014. Submission to the UNFCCC Secretariat, Copenhagen, http://www.umweltbundesamt.de/sites/default/files/medien/384/bilder/dateien/4_tab_thgemi-eu-kyoto-ziele_2014–08–14.pdf, retrieved on 2.3.2015

European Commission, (2005) Communication of 21 September 2005 from the Commission to the Council and the European Parliament – Thematic Strategy on Air Pollution, COM(2005) 446 final, Brussels, 21.9.2005

European Commission. (2006) Guidance Document for the implementation of the European PRTR, http://prtr.ec.europa.eu/docs/EN_E-PRTR_fin.pdf, retrieved on 27.2.2015

European Commission. (2011) Communication from the Commission to the European Parliament ... – A Roadmap for moving to a competitive low carbon economy in 2050 COM (2011) 109 final, Brussels, 15.12.2011

European Environment Agency (EEA 2009), Copenhagen, Denmark, About E-PRTR, http://prtr.ec.europa.eu/pgAbout.aspx, retrieved on 2.03.2013

Eurostat. (06.2012). Wassernutzungsbilanz Deutschland. Retrieved on 15.01.2013 from Eurostat: http://yppsso.eurostat.ec.europa.eu/nui/show.do

Eurostat. (26.02.2013). Bevölkerung am 1. Januar des betreffenden Jahres. Retrieved on 28.05.2013 from European Statistics of the European Commission (Eurostat): http://epp.eurostat.ec.europa.eu/tgm/table.do?tab=table&init=1&language=de&pcode=tps00001&plugin=1

Eurostat. (2014a). Gross Inland Energy Consumption, http://ec.europa.eu/eurostat/tgm/table.do?tab=table&init=1&language=en&pcode=tsdcc320&plugin=1; Retrieved on 3.03.2015 from Eurostat

Eurostat. (2014b). Primärerzeugung von erneuerbarer Energie, http://ec.europa.eu/eurostat/tgm/table.do?tab=table&init=1&language=en&pcode=tsdcc320&plugin=1; Retrieved on 3.03.2015 from Eurostat

Eurostat. (2014c). Statistik Abfallaufkommen 2010, http://epp.eurostat.ec.europa.eu/statistics_explained/index.php/Waste_statistics/de Retrieved on 29.10.2014 from Eurostat

Frischknecht, R., Steiner, R., & Jungbluth, N. (2009). Methode der ökologischen Knappheit – Öko-faktoren 2006. Methode für die Wirkungsabschätzung in Ökobilanzen. Umwelt-Wissen no. 0906. Berne: Swiss Federal Office for the Environment (BAFU).

Frischknecht, R., Büsser Knöpfel, S. (2013). Ökofaktoren Schweiz 2013 gemäß der Methode der ökologischen Knappheit. Methodische Grundlagen und Anwendung auf die Schweiz. Bundesamt für Umwelt, Berne. Umwelt-Wissen No. 1330.

HELCOM. (1988). Declaration on the Protection of the Environment of the Baltic Sea. Helsinki: Baltic Marine Environment Protection Commission (HELCOM).

INC, (1987/1995), Ministerial Declaration for the 2nd and 4th INC. London/Esbjerg: International Conference on the Protection of the North Sea (INC).

IPCC. (2001). Climate Change 2001: The Scientific Basis. Contribution of Working Group I to the Third Assessment Report of the Intergovernmental Panel on Climate Change. (J. Houghton, Y. Ding, D. Griggs, M. Noguer, P. van der Linden, X. Dai, C. Johnson, eds.) Cambridge, UK: Cambridge University Press.

ISO 14040:2006 Environmental management – Life cycle assessment – Principles and framework; ISO Institute

ISO 14040:2006 Environmental management – Life cycle assessment – Requirements and guidelines ISO Institute

Jankiewicz, P., & Krahe, P. (2003). Abflussbilanz und Bilanzierung der Wasserströme. In L. Leibniz-Institut für Länderkunde, Nationalatlas Bundesrepublik Deutschland – Relief, Boden und Wasser (pp. 148–149). Heidelberg Berlin: Spektrum Akademischer Verlag. Retrieved on 28.05.2013 from nationalatlas.de: http://yrchiv.nationalatlas.de/wp-content/yrt_pdf/Band2_148-149_archiv.pdf

OECD. (2008). OECD Key Environmental Indicators. Paris: OECD.

OECD. (2013). Water consumption. In OECD Facebook 2013: Economic, Environmental and Social Statistics. OECD Publishing.

PARCOM. (1988). PARCOM Recommendation 88/2 on the Reduction in Inputs of Nutrients to the Paris Convention Area. Lisbon: OSPAR Commission for the Protection of the Marine Environment of the North-East Atlantic.

Council of the European Union. (15.05.2002). Council Decision (2002/358/EC) of 25 April 2002 concerning the approval of the Kyoto Protocol to the United Nations Framework Convention. Retrieved on 08.01.2013 from eur-lex.europe: http://eur-lex.europa.eu/LexUriServ/LexUriServ.do?uri=OJ:L:2002:130:0001.0001:DE:PDF

UBA. (08 2010d). Wasser, Trinkwasser und Gewässerschutz – Schutz der Meere – Internationale Nordseeschutz-Konferenzen (INK). Retrieved on 16.05.2013 from Federal Environment Agency (UBA): http://www.umweltbundesamt.de/wasser/themen/meere/internationale-nordseeschutz-konferenzen.htm

UBA. (09 2012b). Emissionen der sechs im Kyoto-Protokoll genannten Treibhausgase. Retrieved on 08.01.2013 from Federal Environment Agency – Daten zur Umwelt: http://www.umwelt bundesamt-daten-zur-umwelt.de/umweltdaten/public/theme.do?nodeIdent=2726# Rechtsgrundlagen

United Nations Framework Convention on Climate Change. (09.05.1992). Kyoto Protocol to the United Nations Framework Convention on Climate Change. Retrieved on 04.01.2012 from http://unfccc.int/essential_background/kyoto_protocol/items/1678.php

Printed in the United States
By Bookmasters